Volcanic Eruptions, Tree Rings and Multielemental Chemistry

An Investigation of Dendrochemical Potential for the Absolute Dating of Past Volcanism

Charlotte L. Pearson

BAR International Series 1556
2006

Published in 2019 by
BAR Publishing, Oxford

BAR International Series 1556

Volcanic Eruptions, Tree Rings and Multielemental Chemistry

© Charlotte L. Pearson and the Publisher 2006

The author's moral rights under the 1988 UK Copyright,
Designs and Patents Act are hereby expressly asserted.

All rights reserved. No part of this work may be copied, reproduced, stored, sold, distributed, scanned, saved in any form of digital format or transmitted in any form digitally, without the written permission of the Publisher.

ISBN 9781841717623 paperback
ISBN 9781407330143 e-book

DOI https://doi.org/10.30861/9781841717623

A catalogue record for this book is available from the British Library

This book is available at www.barpublishing.com

BAR Publishing is the trading name of British Archaeological Reports (Oxford) Ltd. British Archaeological Reports was first incorporated in 1974 to publish the BAR Series, International and British. In 1992 Hadrian Books Ltd became part of the BAR group. This volume was originally published by John and Erica Hedges in conjunction with British Archaeological Reports (Oxford) Ltd / Hadrian Books Ltd, the Series principal publisher, in 2006. This present volume is published by BAR Publishing, 2019.

BAR titles are available from:

BAR Publishing
122 Banbury Rd, Oxford, OX2 7BP, UK
EMAIL info@barpublishing.com
PHONE +44 (0)1865 310431
FAX +44 (0)1865 316916
www.barpublishing.com

Acknowledgments

This work took place as part of a three year PhD project at The University of Reading. It was supervised by professors Sturt Manning and Max Coleman with the help of Dr Kym Jarvis and the staff of the NERC ICP facility at Kingston-Upon-Thames. It was funded by a University of Reading studentship and the Natural Environment Research Council. Professors Martin Bell and Paul Buckland were examiners, with Martin also acting as advisory convener. During the course of the project a large cross-section of the dendrochronological community were kind enough to contribute wood samples and advice. Particular thanks go to Mike Baillie and David Brown of the School of Archaeology and Palaeoecology, Queen's University Belfast; Peter Ian Kuniholm and Maryanne Newton of The Malcolm and Carolyn Wiener Laboratory for Aegean and Near Eastern Dendrochronology; Christer Karlsson and Håkan Grudd of the Swedish University of Agricultural Sciences, Siljansfors Experimental Forest; Gretel Boswijk and John Ogden, of the University of Auckland; Þöstur Eysteinsson and Hrafn Oskarsson of the Icelandic Forestry Commission; Mike Barbetti of the Australian Key Centre for Microscopy and Microanalysis, The University of Sydney; Rosanne D'Arrigo and David Frank of the Tree-Ring Laboratory of Lamont-Doherty Earth Observatory, Columbia University; and Martin Bridge of University College London. I'd like to say special thanks to Martin and Sturt, and all my friends for their support throughout. Finally I would like to say a big thank you to my parents, Ken and Vicky Pearson and Tony and Margaret Brewer, and my husband, Peter, for his unwavering support and encouragement.

Contents

1 Introduction   **1**

    1.1 Volcanic eruptions and archaeology 1

    1.2 Dendrochronology and event years 3

    1.3 Aims and objectives 5

2 Dendrochemistry   **7**

    2.1 Introduction 7

    2.2 Basic wood anatomy 7

    2.3 Principles of dendrochemistry 8

    2.4 Database design 12

    2.5 Database outputs: element behaviour in tree ring sequences 13

    2.6 Database outputs: study success 17

    2.7 The dendrochemistry of volcanic eruptions 18

    2.8 Conclusions 21

    2.9 Tables 21

3 Development of Methodology   **36**

    3.1 Field sampling 36

    3.2 Analytical techniques 38

    3.3 Methodological Design: Laser Ablation ICP-MS 40

    3.4 Methodological Design: Solutions ICP-MS and ICP-AES 57

    3.5 Conclusions 62

4 Pilot Studies   65

4.1 Introduction 65

4.2 Pilot study 1 65

4.3 Pilot study 2 69

4.4 Pilot study 3 75

4.5 Pilot study 4 76

4.6 Pilot study 5 84

4.7 Pilot study 6 93

4.8 Conclusions 98

5 Case Study 1 - Sweden   99

5.1 Introduction 99

5.2 Site selection 99

5.3 Field sampling 100

5.4 Laboratory preparation 102

5.5 Environmental signatures to trace 103

5.6 Results 108

5.7 Conclusions 135

5.8 Summary 138

6 Case Study 2 - Porsuk, Turkey   141

6.1 Introduction 141

6.2 Aims 141

6.3 Laboratory preparation 143

6.4 Results 143

6.5 Conclusions 167

6.6 Summary 171

7 Discussion   172

7.1	Introduction	172
7.2	Dendrochemical review	172
7.3	Methodological development	173
7.4	Results	173
7.5	Conclusions	175
7.6	Concluding remarks	176

Bibliography	**177**
A Tree species - common names	189
B Element names and symbols	191
C CEM Digestion of Wood	193
D Set-up for LA-ICP-MS analytical conditions	194
E Method for soils into solutions	195

Chapter 1

Introduction

1.1 Volcanic eruptions and archaeology

Volcanic eruptions are diverse, unique events, occupying a narrow window in both time and space. The impact they have on human society is determined by the individual eruption characteristics, the complexity of any affected society, and the proximity of that society to the eruption source. Over the last century we have seen numerous examples of the range and scope of volcanic impact. The abrupt desolation of St Pierre with the sudden eruption of Montagne Pelée in 1902, and the gradual devastation of the island of Montserrat by the ongoing Soufrière Hills eruptions, (begun in 1995), provide contrasting examples of the dramatic localised impact of volcanism. In terms of wider ranging impact, the opportunity to study major, explosive eruptions like El Chichon in 1982 and, most significantly, Mt Pinatubo in 1991 [100], has done much to further our understanding of the potential repercussions of such events on hemispheric and global climate[1]. Back beyond the last hundred years, far larger eruptions than these are known to have shaken the globe. Key examples of are Krakatau in 1883, and Tambora in 1815. Tambora in particular is recorded to have had worldwide repercussions on climate, (for example see [126]), however, even this catastrophic milestone is dwarfed by other events further back in anthropogenic and global history. From geological, archaeological and climatological records, we now know of *super eruptions*, which some researchers argue to have had such marked effects as to divert the course of human history [128]. For example, the eruption of Toba, 74,000 years ago in Sumatra, may have pushed humanity to the brink of extinction in the wake of a *volcanic winter* that lasted several years [125, 174, 127] [2]. How would the societies of today cope if this scale of event were to occur again? What would happen, for example, in the event of an eruption of the giant caldera at Yellowstone National Park, already 40,000 years overdue [144]? We saw in 1981 how dramatic the effect of the relatively small scale eruption of Mt St. Helen's was on a society dependent on communication and transport systems. How would the modern world cope with a Yellowstone eruption which could potentially be 2,500 times the size of Mt St Helens?

Archaeological (and other paleoenvironmental) research into the impact of volcanic eruptions on past human societies, can help with predictions for future response and the impact of events for which we have no modern analogue. For archaeology itself though, the study of past volcanic eruptions and contemporary human societies can provide a wealth of insight and information, from the nature of impact to the preserved intricacies of material culture.

It seems that ever since the human timescale began, people have been drawn to the richly fertile soils of volcanic foothills, and thus lived in close proximity to some of the most potentially deadly volcanoes in the world. A glance

---
[1] Evidence published in November 2003 connects explosive volcanic forcing with El Nino-like climate events [1]
[2] It should be noted that the extent of the impact of Toba has been disputed [111]

at the massive conurbation which nestles in the shadow of Mt. Vesuvius, in the present day Bay of Naples, illustrates the fact that for the societies of today, as for those in the past, the threat of immense devastation or even loss of life does not deter people from the short term gain. Against the scale of a human life time, the benefits of living on the foothills of a volcano can often outweigh the risks. The occasional times when this gamble has not paid off, have provided the archaeological record with some of its richest and most fascinating sites.

Whilst many kinds of natural disasters detected in the archaeological record may have induced social repercussions of interest to the archaeologist, volcanic eruptions afford a unique opportunity to study a site before and after the event as well as providing the potential to understand change in immense detail and to great effect. The reason for this is that in addition to highly destructive molten lava flows, volcanic eruptions can also deposit great quantities of relatively fine grained debris in a very short space of time. This occurs as the violent release of gas from a volcanic vent causes the fragmentation and expulsion of shattered rock and / or disrupted lava. These *pyroclastic* (from the Greek for *fire* and *broken*) fragments, ejected into the air, range from spindle shaped volcanic bombs (>64mm), to the smaller, sub-spherical lapilli (2-64mm) to ash (<2mm). A broad term for all these deposits which will principally be used in all further description is *tephra*, (from the Greek for ash) [38]. Deposition of pyroclastics can lead to rapid burial of the contemporaneous land surface beneath a protective layer of either very deep or semi-impenetrable strata, thus providing a startling level of archaeological preservation. This astounding degree of preservation has provided a unique insight into the material culture and social attitudes of particular societies prior to (and during) various key eruptions, call to mind, for example, the stricken figures of Pompeii and Herculaneum, or the amazing frescoes of Akrotiri.

However, the importance of tephra fall does not end here. By far the most important consequence of its rapid dispersal and deposition is, that in coating the contemporaneous land surface it forms a stratigraphic marker horizon for a very narrow window in time. The usefulness of this to archaeological and paleoenvironmental research cannot be overstated. It allows for correlation of contemporary sites and events for the whole geographical spread of a particular eruption horizon, and this can be very wide ranging indeed. For example, tephra from the 1259 AD eruption of El Chichon, Mexico, has been found in both the Greenland and Antarctic ice caps [115]. Thus, tephra horizons not only provide a starting point for the analysis of impact on sites obviously effected by a volcanic eruption, but also for sites at much greater distances from the source, which could only have been indirectly impacted by wider ranging, less tangible effects.

Tephrochronology - the use of tephra layers as a chronological tool, was pioneered in Iceland, by Sigurdur Thórarinsson, and has since been used in many studies across the globe, for example [39, 59]. It involves the identification of a given tephra horizon by means of the unique geochemistry, mineralogy, petrography or granulometry of that specific horizon, for example [105, 139]. Once a particular horizon has been characterised, not only can it be distinguished from any other tephra layer, but it may also be possible to identify the source area. From this point the horizon can be used to date associated stratigraphy. On a simple level, this may involve the relative dating of sediment layers overlying and underlying the horizon, but more specific dates can also be attributed by a range of modern scientific dating methods. For example radiocarbon dating can be used on associated organic material [105, 109], and in the case of older deposits, Potassium - Argon [16], or fission track dating [102] can be directly applied. Where a particular tephra layer has fallen over a land surface which is part of an annually accruing system of deposition (such as certain types of lake sediment or, most importantly, glaciers and ice caps), it can be dated stratigraphically in relation to annual increments [60, 172]. In a similar way horizons can be dated via biostratigraphical methods such as pollen analysis, by paleomagnetic correlations, and in a deep ocean context, via oxygen isotope stage boundaries (for further details see [137]).

Whilst the degree of resolution achieved with some of these methods of dating can be very accurate [109], the accuracy of the date is dependent on the constraints of the particular dating method used. Even in the very best scenarios dates of plus or minus several years are quoted. When dealing with a human time frame the importance of tying a date down to a specific year is clear, unfortunately, for the majority of volcanic eruptions of the pre-modern period, absolute dates such as this do not exist. So here lies the crux of the problem, amidst all the potential of tephra horizons for the construction of regional, hemispherical and even global chronologies, and for the rigorous investigation of wide ranging volcanic impact on societies of the past, there is one underlying problem, i.e. the majority of the volcanic eruptions of pre-modern times lack an absolute date. One way to counter this and to complement the existing range of dating techniques used in tephrochronology, is to find a away of connecting volcanic eruptions to a dating method which does offer reliable annual resolution.

1.2 Dendrochronology and event years

'Dendrochronology' comes from the Greek for 'tree' (dendro) and 'knowing the time' (chronology). It is founded on the basic principle that concentric growth layers (tree rings) are formed in the wood of trees on an annual basis, and that the characteristics of these increments reflect changes in conditions at the time of growth. By counting the tree rings, the age of a given tree at the time of felling can be deduced. More critically, by using the technique of 'cross - dating' (developed in the early twentieth century by Andrew E. Douglass), it is possible to match the ring pattern of wood of an unknown age with wood of a known age and so overlap a sequence back in time using successively older timbers. This is based on the premise that all trees in the same area and of the same species should record similar growth conditions (see figure 1.1).

Trees that form distinct, reliable annual rings are found throughout boreal and temperate regions. In mid to high latitudes certain species of tree may live from several hundred to around a thousand years (e.g. *Pinus aristata*[3] [43]), and wood can be preserved in peat or buildings, or by cold, dry conditions. Thus, it has been possible to construct long master chronologies that extend back over most of the Holocene [43], and so called 'floating chronologies' which extend back even further [79]. Such chronologies do not rely on the measurement of a single ring pattern but on the replication of many trees. They are cross-dated for every period to produce an average pattern of year-by-year growth spanning the lives of a series of successively older samples for a particular region. Once a master chronology has been constructed, the ring width measurement pattern of any randomly collected sample from the same area can potentially be matched against the master chronology and given a precise date. Master chronologies have been invaluable to archaeology not only as a means of providing an absolute date for building timbers etc., but also in the calibration of the radiocarbon timescale [8]. Besides this potential for dating there is a wealth of other information to be gained from tree rings [34, 134]. Study of the density and width of various years can be used to make reliable inferences as to climatic conditions during any given period of annual growth, and in this way long chronologies of climatic change have been developed [21, 22, 31, 44, 52, 73, 83, 87]. It is in the context of climate that a point of linkage exists between volcanism and tree rings. As a volcano erupts, silicate micro particles and acidic gasses are released into the stratosphere. Whilst the majority of the particles settle to earth relatively quickly, the gasses build into a sulphur dioxide rich stratospheric aerosol which backscatters incoming solar radiation to result in cooling of the lower troposphere [60]. The study of recent major volcanic eruptions such as Pinatubo (1991) [30, 100, 130], has done much to advance our understanding of the potential global climatic impact of volcanic eruptions. Although the relationship between volcanic volatile emission and climate fluctuation is not a simple one [13], it is clear from the literature that there is potential for eruptions to cause significant change to global climate and for this to be recorded in the tree ring record [11, 20, 33, 48, 80, 135, 136, 169]. However, this association is far from universally agreed upon. Whilst the previous referenced examples contain numerous instances of correlations between known and approximately dated volcanic eruptions and abrupt, short lived anomalies in tree ring patterns (as would be consistent with a response to a sudden, short term climate perturbation), the association has not been irrefutably proven. As Buckland *et al* [23, page 581] put it, "*A first rule of statistics is that the existence of a correlation does not itself prove a causal connection*". Pyle, [123, page 90] sum-

---
[3] A list of common names for tree species can be found in Appendix A

FIGURE 1.1: Principles of crossdating. Tree ring widths are measured from successively older samples, in this example a living tree, timber building, archaeological structure and wood preserved in a bog environment.

marises on the dating of volcanic eruptions with the following, *"Indirect methods [tree rings] are beguiling, being potentially more precise but at the same time highly ambiguous, and should only be treated with the utmost caution."*

How then can this be resolved? How might one go about providing a proven, positive causal connection linking specific volcanic eruptions with specific event years in the dendroclimatological record? A clue lies in the records obtained from the ice cores. Much of the debate on the tree ring / volcano connection over the last thirty years has included dating evidence presented from cores taken from annually accruing polar ice sheets. Each year snow laden with a sample of whatever is in the atmosphere at the time is added to the ice mass, this includes traces of the chemistry of any major volcanic eruption occurring in a particular year. Chemical analysis of the yearly increments can in theory link a particular volcanic event with a specific year. In earlier studies the chemical link was usually an acidity spike resulting from sulphurus emissions [60], however more recently rises in other elements, e.g. bismuth[4] (Bi) [27] or lead (Pb) [133] and the presence of tephra shards [115] have been observed to correlate with the known dates of volcanic eruptions. At best this evidence can not only prove a volcanic origin, but can also identify the source volcano by means of providing a chemical match. The problem is that the dating method itself is not entirely reliable. As White et al. [107, page 86] point out *"When information is viewed on long timescales, ca. 100,000 years, and with a time resolution of decades to centuries, glacial ice comes very close to acting as an ideal recorder. As one looks closer, decreasing the time scale, increasing the time resolution ... the imperfections begin to become apparent..."* This is because the further you go back in time, the deeper the ice is buried and the greater the deformation of the layers, thus annual resolution is not possible beyond the last few hundred years.

So it seems that the dendrochronological record, which is accurate at annual resolution back to the early Holocene in both the USA and Europe, offers the only viable potential to reconstruct an absolute annual resolution chronology of past major climatically effective volcanism. However, as Pyle [123, page 90] writes; *"Unlike ice cores where the composition of the acidity peaks provides a control on their origin ... there is no such information from tree rings."*. Could the answer to this problem be then, as Hughes [68, page 211] suggests, *"...to demonstrate in tree rings an elemental or isotope fingerprint of major explosive eruptions"*? If such a finger print could be isolated in a tree ring, it could then be compared with the chemical signatures from the ice cores, and from geochemical analysis of tephra sherds, in order to once and for all, positively link an absolute date with a specific volcanic eruption.

This investigation will seek to assess the potential of tree rings as time capsules of the environmental chemistry at their time of formation. It will consider the evidence from plant physiology and from studies in the existing discipline of 'dendrochemistry'. It will also present the results of newly conducted experimental work into the development of suitable protocols for the analysis of individual tree rings in order to trace volcanic eruptions. Finally, results of the new analysis will be presented and the overall potential of tree ring chemistry for dating volcanic eruptions will be summarised.

1.3 Aims and objectives

The primary research aim of this project is to investigate the potential of tree ring chemistry for the interpretation of volcanically derived signatures in absolutely dated annual growth increments.

This shall be achieved through three stages of research:

First, through a review of the discipline of dendrochemistry, highlighting the principles and potential problems general to all dendrochemical analysis. This will include the collation of complimentary information extracted from a wide range of existing interdisciplinary dendrochemical studies. From analysis of this information, the optimum procedure for completing

---
[4]This text contains references to a wide range of elements from the periodic table. The first time each element is cited in the text (tables excluded) the full name of the element and the associated symbol (in brackets) will be given. Thereafter, the elemental symbol only will be given. An alphabetically organised table of element symbols and names is given in Appendix B

a successful dendrochemical study will be developed along with a profile of the best wood types and elements for analysis.

Second, a new methodology will be developed and evaluated in terms of the analysis of individual tree rings by Inductively Coupled Plasma Mass Spectrometry (ICP-MS) - the pre-selected method of choice for the project's exploration of dendrochemistry. Two methods of sample induction will be compared and an optimum methodology developed.

Third, new investigations will be made into the chemistry of a range of dated tree ring sequences covering known and speculated volcanic events from a variety of sites around the globe. This investigation will include a series of pilot studies which consider the use of ICP-MS for different wood types taken from trees growing in different environments at different proximities to volcanoes and covering eruptions of known date. It will also consist of two case studies, one which seeks to demonstrate the complexities of any dendrochemical study on modern wood; and one which considers the problems of applying the principles of dendrochemistry to archaeological samples. This final section considers the practical implications for analysis should the overall investigation show that there is real potential to date the volcanic eruptions of prehistory via the chemistry of tree rings.

Chapter 2

Dendrochemistry

2.1 Introduction

This chapter draws on and reviews a wide range of the available literature on dendrochemical studies. After a brief introduction to wood anatomy, it will begin by introducing the principles of the discipline and considering the various complications and unknowns inherent to any dendrochemical study. It will introduce an identified need for the development of a dendrochemistry database to present a summary of what is currently known in terms of the behaviour of specific elements in various tree species. The design and implementation of the database will be discussed, and examples of database query results will be used to summarise various aspects of what is known about tree ring chemistry. Database outputs will also be used to explore the various techniques used in dendrochemical studies over the years, and the potential success associated with using certain techniques, elements and wood types. Finally three hypotheses developed from evidence in the literature will be put forward as the types of potential signatures to look for to tie tree ring chemistry to volcanic eruptions.

2.2 Basic wood anatomy

An awareness of the nature of the various component parts which go to make up the overall structure of a tree, is fundamental to any dendrochemical research. The main aim of this section is to present a brief overview of this complex area of study, with the primary aim of introducing and defining physiological terminology to be used throughout the chapter and the rest of the thesis.

Figure 2.1 shows the basic anatomy of a typical tree. The trunk of any tree is made up of a number of living and non-living layers. The outermost layer is the bark (or cork), which forms a living, protective layer around the tree, that lowers water loss and insulates against cold and heat. Directly beneath the bark is the phloem, this layer transports sugars around the tree, feeding the active cambium layers of the cork and the vascular cambium. The vascular cambium is the key active layer, usually only a cell or so wide, it divides to produce the annual woody increment of an individual growth ring. Beyond the cambium the wood becomes increasingly older through the active sapwood zone, where nutrients are conducted in solution through tube like cells, towards the center of the largely inactive heartwood.

Wood is not a solid homogenous substance, all these individual layers can be further broken down into a large number of cells. The characteristics of these cells are highly variable from species to species, however the broadest of distinctions can be made by dividing tree species into two primary classes. These are the angiosperms and the gymnosperms. These terms refer to plants which reproduce via flowers (angiosperms) or via cones (gymnosperms). Gymnosperm trees are generally more primitive species, often with a more simplistic wood structure, consisting primarily of long tapered cells known as trachids. A single growth ring

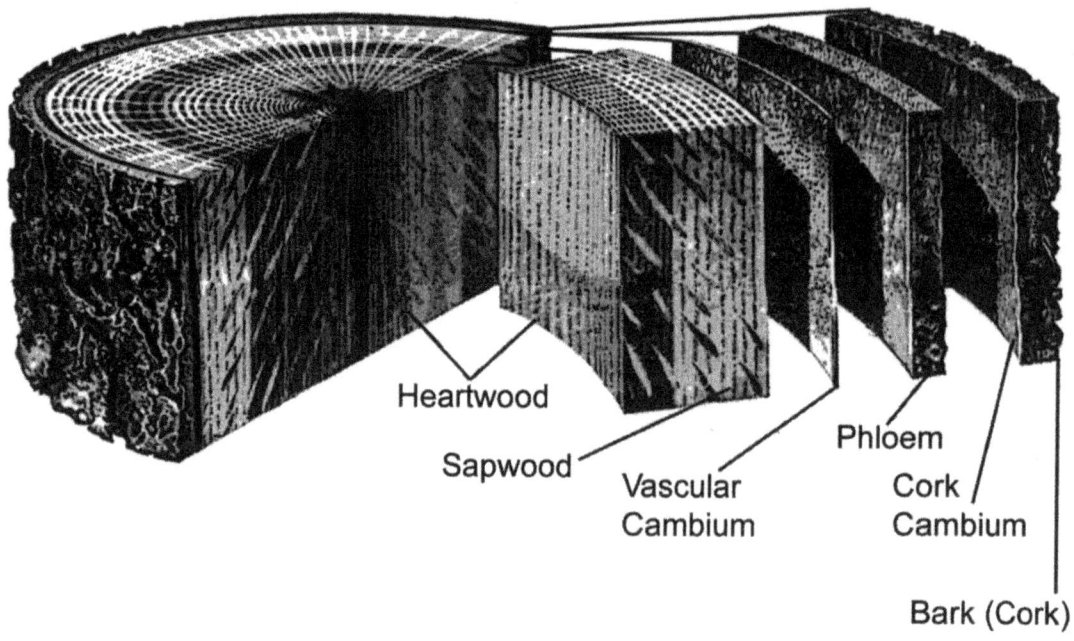

FIGURE 2.1: The basic anatomy of a typical tree.

is defined by large, thin walled trachids of the early wood (spring growth) and smaller thicker walled trachids of the late wood (end of season growth). Angiosperm species have typically more complex structures. They can be diffuse porus like many gymnosperm species, or ring porus (for example *Quercus* sp.). In this case the cells are wide and tabular. A single growth ring is defined by large early wood vessels which gradually decrease with size into the late wood.

Figure 2.2 illustrates a more detailed view of the anatomy of a gymnosperm species. In doing so it introduces additional parts of the anatomy of tree rings, the chemistry of which which will be discussed later in the chapter. Specific elements of the anatomy requiring further explanation include, the epithelial cell walls, the middle lamella, the torus, pit membrane and rays.

Epithelial cell walls - The vertical and horizontal parenchyma cells which line resin cavities, these are built up to form the resin canals which transport resin around the tree.

Middle lamella - A thin layer of intra cellular substance between adjacent cells.

Torus and pit membranes - The part of the characteristic pitting that occurs in various cell walls (especially in coniferous species).

Ray - Part of a sheet of cells running through the wood in a radial direction. Rays transport nutrients around the tree.

2.3 Principles of dendrochemistry

The science of dendrochemistry, has grown up alongside dendrochronology over the past twenty to thirty years, largely with the purpose of tracing histories of anthropogenic pollution. It is based on the principle that "*the chemical make up of the annual woody increment (i.e. the yearly tree ring) at least partly reflects the chemistry of the environment during the year of formation.*" [3, page 1103]. The idea is simple, as a tree puts on its yearly growth, a sample of the chemistry of the growth environment is encapsulated in the tissue and thus, a year by year sequence documenting changes in environmental chemistry is created. In a dendrochemical study this information is accessed by

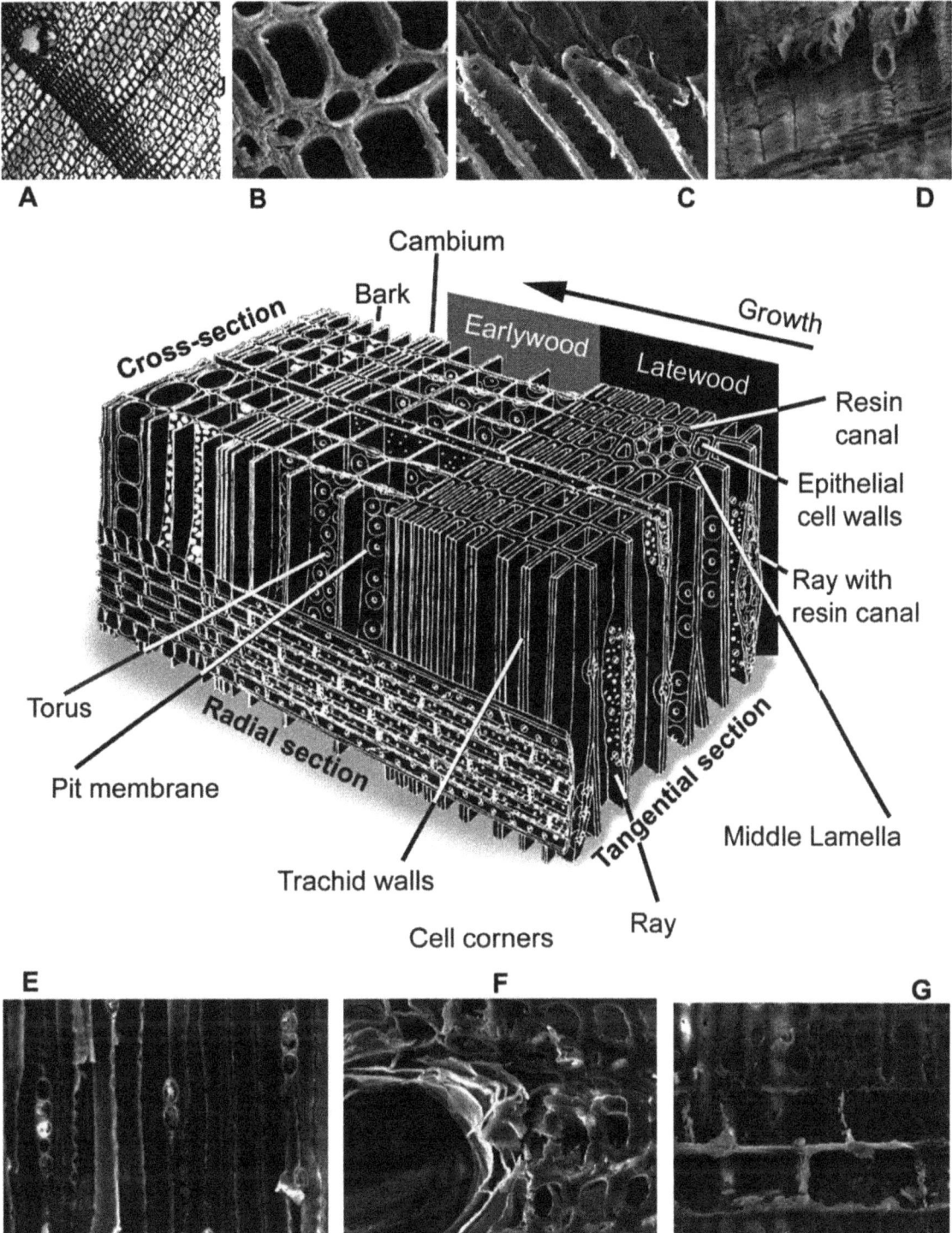

FIGURE 2.2: The anatomy of a gymnosperm, photomicrograph width indicated in brackets: A: (1000μm) Early wood and late wood featuring a resin canal (cross-section). B: (120μm) Cell wall detail featuring cell corners and middle lamella (cross-section). C: (150μm) Tori and pitting on tracheid walls (radial section). D: (200μm) trachids (radial section). E: (350μm) Tangential section featuring rays. F: (160μm) Resin duct in late wood (cross-section). G: (300μm) Radial section featuring tori.

means of chemical analysis of individual rings, or more often, groups of tree rings, within these sequences. In practice, however, this model is impacted by a wide range of variables relating not only to various external factors, but also to the internal physiology of the tree and the individual behaviour of specific elements.

External factors influencing tree ring chemistry encompass all the detail of the growth environment. This can include the nature of the underlying bed rock and soil chemistry, the depth of the substrate, the level of the water table, local wind directions, climate, aspect and slope. All these factors govern not only the background levels of elements available for uptake, but also can do much to contribute to the tree's exposure to various changes in environmental chemistry. A simple example is provided by considering proximity and wind direction. If the wind does not carry emissions from the pollutant source in the direction of a given tree it is unlikely that the tree will register a change in chemistry unless it's growing in very close proximity to the source, or the source is extremely large. If the tree is situated downwind of the pollution source it is far more likely that changes in emission may be recorded whatever the proximity. A more complex example can be provided by considering the nature of the soil in which a particular tree is growing. Studies on the impact of acid rain on various forest systems have shown how changes in soil pH can have a defined effect on elemental uptake by trees [17, 36, 45, 54, 69, 82, 84, 140, 141, 161, 164]. Equally, Cutter and Guyette [32] suggest that the depth of a soil is another important factor. If a soil is shallow it has a relatively high landsurface to volume ratio, thus, soil surface exposure to atmospheric reactants is high in relation to the depth of soil through which they may potentially be dispersed. This means that the soil chemistry is more likely to be altered by atmospheric deposition, temperature, or precipitation. The chemistry of the initial soil type is also very important, especially in response to acidification. If, for example, a tree is growing on a calcareous soil formed over a fast eroding bedrock such as chalk, acidic precipitation may have little effect on the soil and therefore, the tree chemistry. The regular addition of eroding chalk would maintain a certain degree of equilibrium, where increasing environmental acidity would be continuously neutralised. On the other hand, if a tree is growing on a slow eroding, igneous bedrock, the addition of that material to the soil may be very slow. In this case the soil can more easily be brought out of equilibrium, with atmospheric constituents playing a more important role in determining the soil, and thus, tree chemistry.

The nature of the soil and these other factors of external influence, can directly determine the dominant path through which a tree takes in nutrients and therefore takes a sample of its environmental chemistry. Watmough and Hutchinson, [164] discuss how soil acidity can have just such an effect. They suggest that trees growing in more acidic soils with higher metal concentrations, will be dominated by uptake through the roots. Trees growing in less acidic soils with lower metal concentrations, are thought to uptake more directly from the environment through the bark and possibly the leaves. Equally it could be hypothesised that trees growing in deep nutritious soils would dominantly absorb nutrients via the roots, whereas those growing in thin, poor quality soils would be likely to rely on uptake from the leaves and bark. The question of which uptake pathway dominates is of central importance to shaping the responsiveness of the tree to chemical change. Uptake through the roots, bark and leaves will now be considered in detail.

As a tree grows it uptakes compounds in solution via its roots from the surrounding soil. As previously considered, levels of various elements in the soils are controlled by factors such as the underlying bedrock, local vegetation, biological activity, air-borne particulate matter and precipitation. This pathway is generally considered to be the main means of uptake for the majority of elements [67]. From the roots, elements can be transported through many of the xylem vessels, including along the rays towards the sapwood, so it could logically be argued that elements taken up in this way are likely to be dispersed throughout the sapwood (the ten or so years of *living* wood). This would result in a great loss of sensitivity for recording environmental change. Even if, for a particular species, it could be proven that certain elements taken up by the roots are only deposited in the active growth ring, there is still the issue of time-lag and chemical changes occurring in the soil which would effect the representative sensitivity in the same way. Therefore it can logically be assumed that the most successful dendrochemical histories can be produced when trees are growing in an environment where the tree's main uptake mechanisms are via a more direct route (through the leaves or bark), or where the element of interest is of low availability in the soil (for example Pb [160]). According to Watmough [160], elements absorbed by foliage can be deposited in the most recently formed outer tree ring following translocation in the phloem. This has been demonstrated by Lin et al. [88] for manganese (Mn) and zinc (Zn). They applied a solution of radioisotopes of these two elements direct to the foliage of a particular tree, in a single large dose (approx. 12 μg ml), and found that 10% of them were translocated away from the application area and into the new growth ring. As Watmough [160] points out, what may be true of that study may not be true if a different element is selected in different quantities or applications, or if a different tree is used. However, it does provide experimental evidence that in some cases at least, uptake through the leaves has the potential to accurately record changes in environmental chemistry.

The function of the bark as a protective layer shielding the growth area of the cambium makes it seem a less likely pathway for elemental uptake, however this does not in fact seem to be the case. It would appear that the bark, like the leaves, may provide a more direct uptake path for elements than through the roots. Lepp and Dollard [86] applied ^{210}Pb to the bark of several tree species and reported an uptake in both dormant and active wood (indicating that metal absorption through bark can occur on an all year round basis). It should however be noted that Ward et al. [158] suggest that Pb is most likely to enter trees through the roots as they found a sharp decrease in Pb concentrations between the outer and inner bark instead of the steady gradient you might expect if the element was diffusing from the outside in. Watmough [160] however also concludes that uptake through bark is a major pathway by which mobile and immobile elements can become incorporated in tree rings.

Whilst it seems that external influences on the chemistry of tree rings can have complex consequences for the subsequent chemistry of the individual tree rings, they can at least be partially controlled and explained by the selection of suitable growth environments. However, the further complication for dendrochemical studies is that, as Smith and Shortle [143] point out, *"trees are not passive recorders of the external environment. Biological processes mediate the formation of the chemical record."* [page 626]. Trees are complex, living structures which primarily absorb elements to meet specific requirements to sustain growth and maintain health, as well as absorbing a more passive sample of their growth environment. In addition to this, the various elements taken up all have certain characteristics which determine the way in which the tree is able to utilise them.

Whilst many of these internal, physiological influences are poorly understood, several studies exist which have helped to identify some of the specific factors which influence the degree of element mobility, and consequently dispersal in the xylem. These include; ion solubility, the charge to ionic radius ratio, sapwood to heartwood equilibrium concentrations and the essential or non-essential nature of a given element. The lower the mobility, the greater the reliability of a given element in the creation of a chronology of environmental chemistry.

The solubility of the various ions, not only determined by the characteristics of individual elements, but also by factors such as the pH of the sap, governs their concentration in the xylem solution. The more soluble an element, the more mobile it is. The more mobile, the less likely it is to remain in the outermost tree ring and accurately record chemical changes as they occur.

The ionic radius of an element is determined by its mass and electron configuration. The smaller the radius relative to the charge, the stronger the ions will be held to exchange sites on the cell wall. So ions with large radii relative to their charge are more mobile.

At the transition zone between the sapwood and the heartwood, differences in the equilibrium of specific element concentrations can govern whether new heartwood is enriched or depleted in a particular element. Cutter and Guyette [32] illustrate this with the example of calcium (Ca). If Ca concentration is high in the sapwood in relation to the mobile fraction of Ca on exchange sites and in insoluble crystals within the transition zone, then it will be stripped from the transition zone at the time of heartwood formation and cycled into the sapwood. In the same paper it is also suggested that a decrease in starches and sugars can result in the formation of phenolic compounds which can yield concentrations of extraneous elements at the boundary.

Finally, elements essential to the growth of the tree, such as (copper) Cu, Ca, Zn, magnesium (Mg), iron (Fe) and K for example, may be preferentially translocated around the tree via specially developed methods such as chelation. This results not only in increased mobility of some less mobile elements, but also in preferential distribution of these elements around the tree. Such elements must therefore be approached with caution in dendrochemical studies, but can also prove very useful in that trees will often take up as much as is available, and thus variations may accurately reflect external chemical change.

These types of influences have led to the occurrence of certain documented, characteristic patterns for particular elements in various species of tree. However, in reviewing the literature it became clear that whilst numerous disjunct studies of various sites, species and elements exist, there is no one reference which summarises profiles for the known behavior of elements in certain species. A database was designed with the primary aim of producing a readily accessible data source which could be used to aid in the interpretation of subsequently analysed data sets. The idea was to produce a bank of data which could be added to throughout the course of the project, and searched for fast reference to the range of information available from the literature.

2.4 Database design

For databases to work most effectively, an efficient structure must first be designed. The first stage of the design process was to assess what data was going to be stored within the database. The four main 'objects' identified were: observation; reference; taxon and element. These objects form the primary tables within the database.

Associated with these objects are a number of 'properties'. For instance, each reference has, amongst other things, one or more authors, a date of publication and a title. The fields associated with the reference, taxon and element tables are as follows:

Reference - reference type; author; editor; year; title; chapter title; journal; volume; pages; publisher; publisher address; dewey number.

Taxon - genus; species; family; vernacular name; full name; authority; Angiosperm/Gymnosperm

Element - element symbol; element name; atomic number; atomic mass; essential element.

The fields for the observation table required considerably more thought. Through initial examination of the literature a number of key observations were identified, with the final selection of fields for the table as follows:

Technique - the analytical technique used. This field offers a selection of techniques supplied by an additional 'dictionary' table. This enables newly discovered techniques to be added to the database without alteration to the database structure.

Associated anatomy - whether a given element is associated with a particular feature of the wood anatomy. This includes a pick list of early wood, late wood, cell corners, epithelial cell walls, resin canal, middle lamella, trachid walls, torus, pit membranes, ray, pith, bark [1] and isolated patches of one or more elements.[2]

Pith to bark pattern - whether a given element shows a general trend from pith to bark. This features a pick list of 'increase, decrease, stable'.

Heartwood/sapwood boundary pattern - whether the concentration of a given element has been shown to alter at the boundary between the heartwood and the sapwood. This features a pick list of 'increase, decrease, peak'.

---

[1] Figure 2.2 illustrates these anatomical features.

[2] Isolated patches of various highly concentrated elements have been recorded to occur within individual tree rings (see section 2.5.4 for further details). These have no association with a particular part of the anatomy but have been recorded in this section as worthy of further consideration.

Xylem mobility - how mobile a given element has been shown to be in the xylem. This features a pick list of 'high, low, moderate'.

Response to acidity - whether the concentration of a given element in wood appears to respond to changes in environmental acidity. This features a pick list of 'increase, decrease, stable'.

Success of element detection - whether a specific element has been detected in the wood via a particular technique. This features a tick box.

Success of study - whether a pollution history has been detected in tree rings and positively correlated with a known pollution event. This features a tick box.

Average concentration - the average concentration found for a particular element from a sequence of tree rings.

Notes - this field allows for the addition of extra miscellaneous information which requires recording, but does not fit within a specifically labelled field.

The four primary and two dictionary tables are related by a number of database joins into a relational database structure. The entity relationship diagram for the database is illustrated in figure 2.3. These relationships enable querying of the database in a wide variety of ways. This produces new insights into the jumble of information located within the literature which would otherwise have gone un-noticed.

In order to facilitate the regular addition of new records throughout the course of the project as new papers were found, an interface was designed in Microsoft Access. Figure 2.4 shows examples of the two forms designed for the reference and observation tables.

This database remains a work in progress to which new additions were made throughout the project and which continues to be updated with the aim of publication. So far it compiles a total of 651 observations from 79 different species of tree collated from 208 papers.

The potential power of the database is best explained by selecting a random question as an illustration. For example, having conducted chemical analysis on a sequence of tree rings, it might be found that an increase in copper concentrations occurred from the pith to bark. If one wanted to know if such a pattern had been reported by anyone else, i.e. 'Who has shown that copper can increase from pith to bark?' a simple query could be designed as follows. The key parts of the question relate to the reference table, the element table and the observations table. From the reference table the author and year field are selected. From the element table the copper field is selected and from the observation table the pith to bark field is selected. This query results in an output which shows all pith to bark patterns and can be simply modified to show the increasing pattern only, thus answering the question. The same question could be asked for any other element in the periodic table, for a decrease pith to bark or for a change at the heartwood - sapwood boundary, and for any number of other questions.

2.5 Database outputs: element behaviour in tree ring sequences

Observed patterns of element behavior within tree ring sequences fall into three main categories, these are: changes associated with the boundary between the heartwood and the sapwood, overall radial distribution patterns which decline, increase or are stable, and associations of specific elements within the anatomy of a single tree ring. These patterns and occurrences will be discussed in turn and the database outputs will be presented to show how associations occur in particular wood types for particular elements.

2.5.1 Change at the heartwood - sapwood boundary

The boundary between the heartwood and the sapwood is essentially the border between the non-active and active parts of the tree. It can be marked by visually observable changes in the density and colour of the wood, or by decreases in levels of moisture and permeability. The inner heartwood, is the largely mechanically and

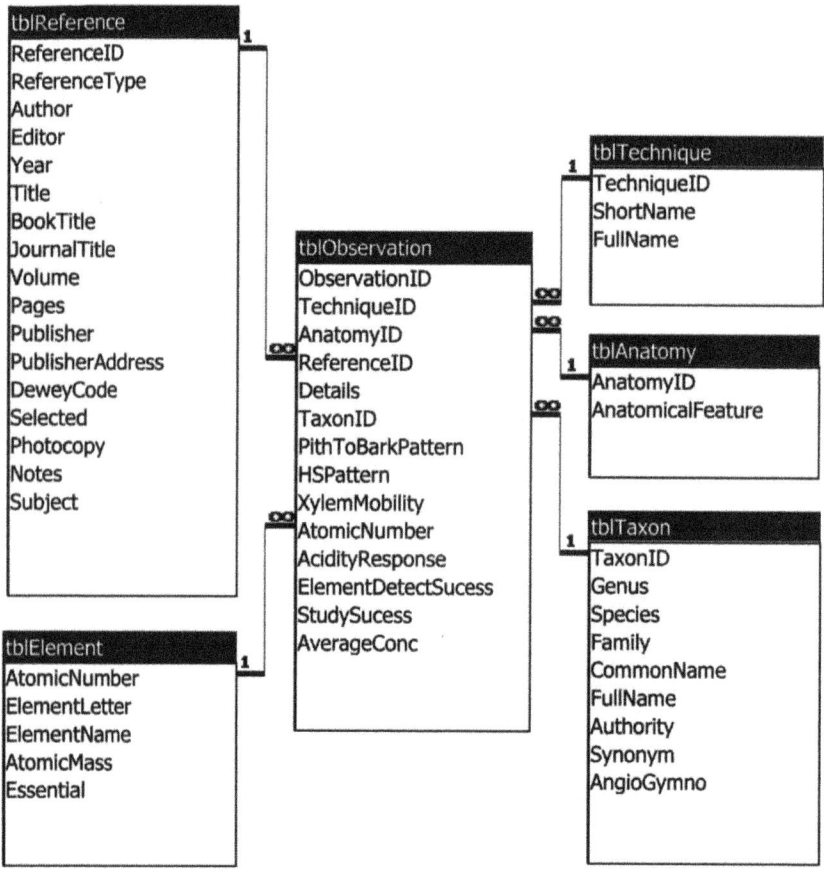

FIGURE 2.3: An entity relationship diagram to show how the connections are made between related fields to link the tables together to form the complete structure of the relational database. The infinity symbol denotes where many component parts may be related to 1.

physiologically non-functional core. The sapwood is considered to be the physiologically active part of the tree and usually consists of the twenty to thirty outermost rings (although this number varies greatly within and between species and with the age of a given tree). The functions of the heartwood and the sapwood are underlain by obvious considerations for the dendrochemist. According to Stewart [148] the heartwood provides a dumping ground for waste products which are used to increase the mechanical strength of the tree. The implication of this is that the elemental content of the heartwood may be governed more by the internal requirements of the tree, than by changes in the external chemistry. As the oldest sapwood rings become incorporated into the heartwood, it may be that the initial chemistry of the tree ring is masked or even totally erased. It could, however, be that this only applies to specific elements and the majority of the signature is preserved. Equally, the heartwood formation process may make a uniform contribution to the chemistry, against which original fluctuations can still be measured. As the main function of the sapwood is to transport nutrients around the tree, a possible problem exists in terms of the potential resolution of histories of chemical change. It may be that lateral movement of sap within the sapwood zone can lead to the diffusion or translocation of various elements between rings. If this was the case, there could be a smearing effect, whereby chemical signatures for particular years could

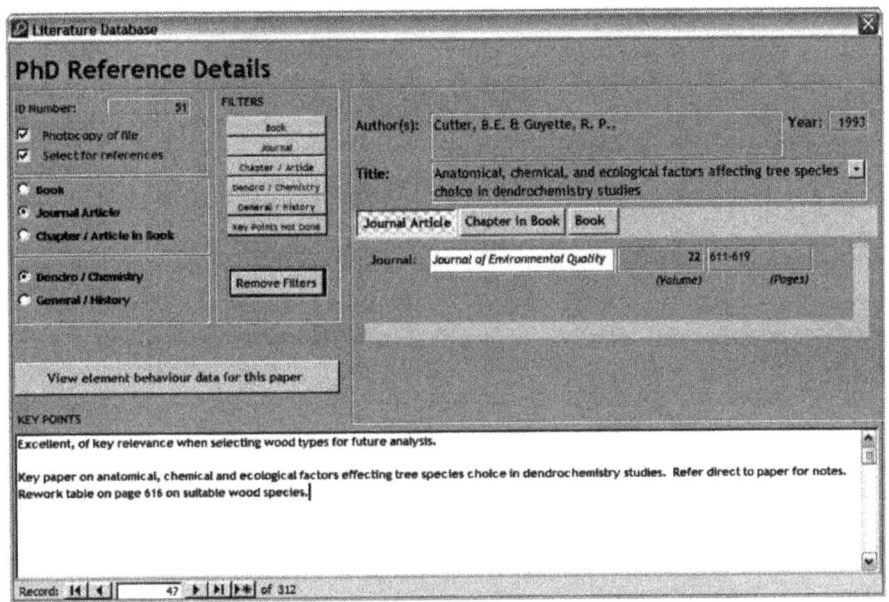

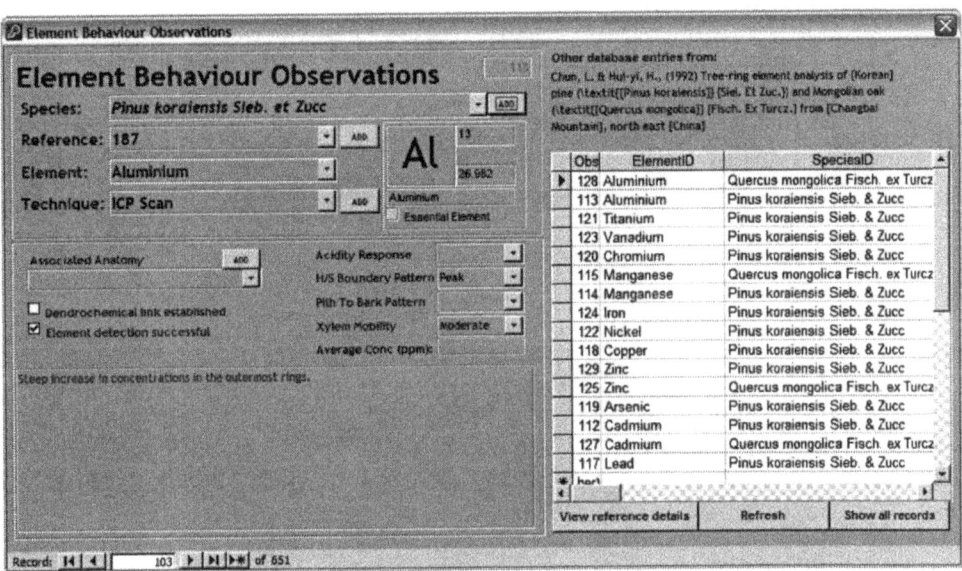

FIGURE 2.4: Screen shots of the form interfaces designed to aid data entry for A, the references table and B, the observations table.

be smudged over several years. This type of effect would presumably be limited to particular elements and would differ between tree species. Watmough [160] cautions that this has not yet been measured under field conditions. However, according to Häsänen and Huttunen [63], and Baes and McLaughlin [5] for some species at least, there is no significant transfer of elements from one annual ring to another.

So how real either of these possible problems are remains to be seen, and in all likelihood will be down to the behaviour of individual elements in individual wood types. Tables 2.1 to 2.3 (see pages 21 to 23) show known element and species associations which have shown significant differences in composition between the heartwood and sapwood in the literature. One of the key papers entered into the database in this regard, is that of Okada et al. [108]. They were the first to identify the three types of pattern into which the tables are divided. These are increases, decreases and peaks of certain elements at the boundary. Where element content was found to increase outwards across the heartwood - sapwood boundary, it was taken as an indication that the elements were removed from the sapwood tissue as the heartwood forms. Where element content was found to decrease outwards across the heartwood - sapwood boundary it was taken as an indication that the elements were accumulated in the tissue during heartwood formation and stayed intact. Thus, an apparent decline was produced on entering the sapwood (this was found to mostly apply to the alkaline earth metals and halogens). Lastly, they found some elements which showed a peak at the heartwood - sapwood boundary which they took to indicate that the element content showed a temporary increase during heartwood formation. Their paper shows how these patterns varied for the same element in different tree species, but also, how in some species all elements were distributed similarly.

Elements detected as both increasing and decreasing in studies of different angiosperms were barium (Ba), Ca, Cu, K, Mg, Mn, sodium (Na), phosphorus (P), strontium (Sr) and Zn. However only Na was found to both increase and decrease in the same tree. No gymnosperms were found to increase and a mixture of both were found to peak at the heartwood - sapwood boundary. Whilst this information is interesting, it also illustrates the dangers of drawing conclusions from database outputs. The potential bias derived from the number and type of records entered must always be considered. For example, in this case the information that no gymnosperms were found to have increasing levels of elements in the sapwood could mislead the reader into concluding that this is the actual case for gymnosperm species. In in reality, it may simply be the fact that most studies have been carried out on angiosperms. This does not undermine the power of the database, but simply illustrates the type of caution needed when attempting to interpret the data in this way. The tables included in this chapter serve to illustrate a sample of what is currently known. The main design function of the database is underlined as being to search for specific facts to aid the interpretation of newly attained data sets.

2.5.2 Radial distribution patterns

Some studies, for example Latimer et al. [81] report no abrupt shifts at the heartwood - sapwood boundary. However, several studies have observed more general trends in the overall patterns of elements from the pith to the bark. For example, Hagemeyer et al. [57] found distinct radial distribution patterns for cadmium (Cd), Pb and Zn, increasing and decreasing in beech trees. The patterns for these and other elements relative to various wood types are found on tables 2.4 to 2.6 (see pages 24 to 26).

Further database queries show examples of Mg, K, Mn, and Cu increasing and decreasing from pith to bark in angiosperm and gymnosperm species. Of the studies completed only aluminum (Al), Ca, carbon (C), Zn, Cd, Ba and Pb have been shown to both increase and decrease in gymnosperm species.

2.5.3 Concentrations of various elements

The concentrations at which particular elements occur in various tree species are highly variable, both between and within the same species grown on different sites [46]. Table 2.7 (page 27) provides examples of the range of average concentrations detected in various wood types via various techniques.

2.5.4 Differences within individual tree rings

The other type of elemental patterns observed in a number of studies, were associations of particular elements within individual tree rings. These have been reported to occur either as highly concentrated clusters with a seemingly random pattern, or are directly associated with anatomical parts of the wood structure. MacLauchlin *et al.* [93], McClenahen and Vimmerstedt [99], Lövestam *et al.* [90] and Sunden *et al.* [150] all report concentrated trace metal accumulation in small areas within individual rings. Sunden *et al.* refer to these clusters as *hotspots* and report patches of Fe, Zn, Cu which were twenty times the concentration intensity of the surrounding trachid walls. Watmough and Hutchinson [163] suggest this effect could be due to sudden, large scale mobilisation of divalent cations due to acidic deposition on soils.

In addition to this, other studies (for example Hoffmann *et al.* [65]) have reported marked differences between the composition of the spring grown early wood and the summer to autumn late wood of the same year.

High concentrations of most elements have also been reported (Bailey and Reeve [6] and Saka and Goring [132]) in the tori and pit-membrane regions of individual tree rings as well as in association with rays and resin ducts and other anatomical features within individual tree rings. A summary of the databased information for these element / anatomy relationships is given on table 2.8 (page 28).

2.6 Database outputs: study success

2.6.1 Analytical techniques

Over the years a wide range of analytical techniques have been used in dendrochemical studies. Table 2.9 (page 29) provides a list of the range of analytical techniques used in dendrochemical analysis. Of these, the most widely used in studies which successfully traced the onset and cessation of pollution were Atomic Absorption Spectroscopy, Inductively Coupled Plasma Atomic Emission Spectroscopy, and Inductively Coupled Plasma Mass Spectrometry techniques. The latter are the relative newcomers to the field of research, having only been used from around 1990 onwards, and as can be noted from table 2.10 (page 29), they can be used to scan for a wide range of elements.

The selection of an analytical technique for a particular study, depends on a number of variables such as the degree of resolution required and, most notably, the elements of interest. Table 2.10 (page 29) lists elements, along with the various analytical techniques which have been used to detect them in dendrochemical studies. This table can be used as an aid to the selection of a certain technique to detect a particular element of interest. However, it should be noted that this information may be strongly biased by which elements have been of interest in other studies, rather than reflecting the true capacity of each technique to detect specific elements. All that can truly be said of the table is that certain techniques can be used to detect certain elements.

2.6.2 Species suitability

Table 2.11 (page 30) lists the tree species used in the dendrochemical studies included in the database. Out of this wide range of species, *Pinus* sp., *Picea* sp., *Quercus* sp. and *Acer* sp. were included in the largest number of studies. A positive dendrochemical link was established for 37% of *Pinus* sp. studied, 46% of *Quercus* sp. and 60% of *Acer* sp. Only 20% of studies for *Picea* sp. produced a positive dendrochemical link. Some examples of species found to be specifically unreliable were; *Fagus sylvatica* [57, 101], *Quercus prinus* [36] and *Liriodendron tulipifera* [99]. From this, one could begin to suggest which species might or might not therefore be the best from which to select samples for future dendrochemical analysis. However, in order to more fully understand the pros and cons of the use of various wood types, and to aid in the selection of the very best species for analysis, a more detailed examination of the literature is required.

At present, research is split on whether primitive gymnosperm species (e.g. *Pinus* sp. [32, 70]) or more complex angiosperm species (e.g. *Acer* sp. [163, 167, 165]) will prove the more successful for dendrochemical analysis. It seems more likely however, that the simple, uniform structure of gymnosperm species may produce better results in terms of a correspondingly simple chemistry against which to read anomalous background input. Legge *et al.* [84] and Zayed *et al.* [171] support this view, suggesting that conifers are the best trees for analysis as they have a primitive structure with few, short ray cells which reduce lateral movement from sapwood to heartwood. In addition to the simple internal structure, gymnosperm species have needles as opposed to leaves. Needles have a far greater surface to volume ratio than flat leaves which potentially results in a greater exposure to atmospheric deposition. This could hypothetically heighten the direct uptake of atmospheric deposition and thus the rapidity of elemental addition to the annual rings. This would mean that such trees may provide a more accurate record of environmental deposition. There is also the fact that gymnosperm species are common, with many species growing in a wide range of environments. Many of them colonise marginal areas to become the dominant vegetation type. The advantages of using a species with such a high ecological amplitude [47] is that comparisons can be made between the wood chemistry from trees of the same type, growing in different conditions and at different proximity's to possible pollutant sources. Whatever the chosen pollutant source, it is likely that there will be sufficient gymnosperms species available to conduct a representative, comparative analysis.

Some authors however, maintain that angiosperms are the more suitable for dendrochemical analysis. For example, Hagemeyer *et al.* [57] point out that in ring porous wood (e.g. *Quercus* sp.) xylem rings loose their ability to conduct water readily a few years after formation. This means that any dendrochemical history constructed from such a tree would potentially be precise to within a couple of years.

In reality, it seems that in choosing a suitable species for dendrochemical analysis one must adopt an holistic approach. Careful consideration must be given to all of the previously discussed factors affecting elemental uptake and distribution in wood, along with a host of site specific factors, as well as evidence from the literature as to the potentially more usable species.

2.6.3 Suitable elements

The degree of success with which certain elements can potentially be linked to specific pollution events, is also controlled by the nature of the element of interest and the way in which it is utilised by a specific species of tree. Tables 2.12 to 2.14 (pages 31 to 32) provide examples from the literature of elements which have been reported to have various degrees of mobility within the tree rings of various species of tree. It should follow that the lower the mobility of the element in a particular species, the greater the potential to trace accurately a change in environmental chemistry related to that element.

The three tables underline the complexity of the situation. It is clear that the mobility of various elements varies greatly with species. For example Mg, which is highly mobile in *Picea*, can be of low mobility in other species such as *Pinus*. However, one could tentatively suggest, based on the data presented in table 2.14 (page 32) that (database bias allowing) *Pinus* would seem to be the species of choice for dendrochemical analysis, as many elements in the tested species were of low mobility. Elements which have been successfully linked to deposition from specific sources are shown in table 2.15 (page 33).

Once again this table is biased in favour of pollutant elements such as Pb which have been of the greatest interest in previous studies. However it does show, that the elements listed, in the right species of tree, under the right conditions, can be successfully used to trace histories of environmental change.

2.7 The dendrochemistry of volcanic eruptions

The final step in investigating the existing dendrochemical literature was to consider how the various studies could be used to support a hypothesis linking tree ring chemistry with volcanic eruptions. Irrespective of the various complications involved with dendrochemical studies, the literature has provided many examples of how the onset and cessation of anthropogenic pollution can be successfully traced via tree ring sequences. It has also shown how specific elements can be used to relate years to events. So, even though the degree of accuracy in terms of time scale can be variable, it does seem that real potential exists not only to link a volcanic eruption with an absolutely dated tree ring, but to match the tree ring chemistry with the eruption chemistry to provide the much needed, indisputable link between a date and a specific event.

If we regard a volcanic eruption as a large scale pollution source, significant parallels can be drawn from anthropogenic pollution histories in terms of exploring the main research question. From consideration of a range of dendrochemical studies, (in particular on the impact of acid rain - a common byproduct of both anthropogenic and volcanic pollution), two main hypotheses were produced, predicting the type of elemental signatures one might hope to find in tree rings responding to an eruption.

The first hypothesis relates to the potential impact of a volcanically induced increase in global or local environmental acidity levels. Based on the existing literature, this could increase or decrease the uptake of various naturally occurring elements into a tree. This may occur either by alteration of the soil chemistry, or as a direct result of acidic deposition on the bark and leaves.

A number of papers (see table 2.16 on page 34) for some examples) have successfully shown how an increase in soil acidity, due to increased precipitation of sulphuric or nitric acid from anthropogenic pollution sources, can lead to alterations in the relative availability of nutrients and ions in the soil. This in turn determines which nutrients are taken up and translocated in the wood of the tree, which acts as an ion exchange column with respect to certain cations. Mobilization of cations by an increase in sulphate and nitrate concentrations in the soil solution, should therefore be reflected in the stem wood of certain years as an increase in the concentration of specific divalent cations. Where the effect on the soil is prolonged or the original soil chemistry is more susceptible to leaching, a corresponding decrease in the availability of certain elements may occur.

One drawback with searching for elemental changes in trees associated with this type of response, is that a time lag is likely to occur between the true onset of pollution and the point at which it shows up in the growth rings. For example, Legge *et al.* [84] found a time lag of ten years between the opening of a sulphur (S) recovery plant and changes in elemental uptake patterns. However, extent of the time lag is dependant on the degree of acidic pollution, the natural buffering capacity of the soil and the particular species of tree. For the majority of studies examined, a far more rapid response was recorded. For example, Bondietti *et al.* [17] show how concentration ratios of Al to Ca, Mg and other divalent cations increased rapidly with a contemporaneous rise in sulphur emissions. Shortle *et al.* [141] illustrate that effects can be not only rapid, but very widespread. In their study Ca and Mg levels were high for the same period of pollution for nine forest sites across the north-eastern USA. However, it is likely a more reliable response might be observed in trees that have a stronger dependency on direct uptake through the bark and leaves, than from the soil. Acid precipitation has been shown to enhance susceptibility to adsorption through the bark and leaves of a tree [70, 88, 119], and so, in a similar way to the impact of acidic deposition on soil, one might expect an increase in the concentrations of certain naturally occurring elements at the time of an eruption. Table 2.16 (page 34) provides examples of responses of certain element concentrations in wood to documented increases in environmental acidity.

Whilst the type of associations suggested by the first hypothesis would be the first step in connecting dated years with eruptions, the evidence could still be argued to be merely another proxy indicator. An increase or decrease of a certain element generated in this way could have numerous other explanations and does not prove a connection. The second hypothesis is more specific, and seeks to find a positive, provable causal connection between a dated tree ring and a particular volcanic eruption. It is based on the idea that an excess of a certain combination of elements is expelled into the atmosphere at times of major volcanic eruptions (either in the gaseous cloud or as part of the physical loading of tephra). The hypothesis is simply that a sample of some aspect of the eruption chemistry may be taken up and incorporated as anomalous spikes within individual tree rings. By linking particularly unusual elements, combinations of elements and or higher concentrations of particular elements known to be associated with a certain eruption, an actual chemical finger print could be provided which could be used to positively link volcanic eruptions with dated tree rings.

There are two studies which go some way towards supporting this hypothesis. First, Tendel and Wolf [153] have found that increased environmental sulphur levels can be detected directly by trees. For their study, they selected a site with a poor, sandy soil in order to maximise the potential sensitivity to a change in anthropogenic sulphur emissions. They found that two different tree species growing in widely distant regions under extremely different growth conditions both showed an increase of sulphur in their annual rings in response to recorded emissions of SO_2. Secondly, Hall et al. [58] sampled *Pseudotsuga menziesii* growing fifteen kilometers northeast of the Mount St. Helens volcano. Their analysis showed that two anomalous elemental peaks occurred at AD 1478 and 1490 - dates which closely correlate with past eruptions of the volcano. The anomalies showed concentrations of rare earth elements; Cerium (Ce), neodymium (Nd), lanthanum (La), samarium (Sm), gadolinium (Gd), lutetium (Lu), and thulium (Tm) which could be positively linked with the eruptions. In this case the studied trees had received an estimated fall out of fifteen centimeters of volcanic ash. The uptake of unusual chemical traces expelled by individual eruptions might be increased by the contemporary release of acidic compounds in the eruption cloud, as considered in the previous two hypotheses.

A further paper, published towards the end of 2002, must finally be considered in reviewing how well either hypotheses might work. Padilla and Anderson [114], like Hall et al. [58], used solution ICP-MS. They carried out a study on a single, three hundred and fifty year old *Pinus ponderosa*. Their aim was to conduct a general study to see what traces of the American industrial revolution, acid rain, regional agriculture and the introduction of the motor car were contained within this long record. They found that most regional, local, small-scale environmental changes did not seem to impact elemental concentrations within the tree. However, what they did find, was a series of rises or peaks in the concentrations of Ba, Cu and Zn which they suggest are derived from the eruptions of Laki (1783), Tambora (1815) and Krakatau (1883). The most convincing of these associations are Ba, Cu and Zn for Tambora and Cu for Krakatau. The claimed Laki association is less convincing as the date also coincides with a forest fire which could be equally responsible. The Ba peak for Tambora is largely convincing except that the resolution is somewhat hazy. The whole sequence was sampled at a resolution of five to ten years, and these increments are not all labelled on the given graphs. The five to ten year increment that shows the short term increase in Ba concentrations is the middle one of three unlabelled increments lying between 1806-1810 and 1827-1835. The Cu and Zn anomalies are also quite convincing for Tambora, occurring at about the right time and with the type of abruptness one would hypothesise as a response to such an event. The Cu increase for 1877-1881 is well defined and seems likely, as they suggest, to be associated with the eruption of Krakatau. The explanation put forward to link these anomalies with the specific volcanic eruptions is essentially the same as the first hypothesis discussed in this section. They postulate that acid rain caused by the eruptions decreased soil pH, increasing the solubility of certain elements and plant uptake to create higher element distribution patterns in the tree.

These results are very encouraging in terms of the overall prospects for the success of the research question, although the publication of this data came rather too late on to make much of a contribution to the experimental design for this particular thesis. It should also be considered that, as with Hall et al. [58], these results were not replicated from other trees from the same growth area, and although Padilla and Anderson [114] do attempt to replicate samples from the same tree, they only replicate 10% of those run and they do so with samples taken from the same level within the tree (as oppose to exploring different heights in the trunk). Given the possible complexities of tree ring chemistry discussed in this chapter, thorough replication of complete sequences of tree rings from several heights and orientations in the trunk seems the essential component to any comprehensive study. However, if observed associations such as these can be proved by similar studies, in terms of modern analogues such as Tambora, the prospects for developing this technology to apply to eruptions of uncertain date are greatly improved. Whilst this would not provide the required, undisputable connection between absolutely dated tree rings and volcanic events, it would add a further dendrochronological proxy to frost ring associations [80], as well as providing a more accurately dated parallel to the acidity record from the ice cores [85].

2.8 Conclusions

Before commencing methodological design, an extensive search and in depth review was made of all the possibly relevant literature. From this it became clear that a thorough appreciation and understanding of all the likely complexities of dendrochemical research was required before commencing any study. With this in mind, the need for a database of reported observations was identified, with the aim of facilitating interpretation of results generated from any new analysis. The database structure was conceptualised and data entered throughout the course of this research and beyond.

The range of literature emphasised the complexity of the discipline, but also provided evidence to suggest that there *are* prospects for developing a dendrochemical resolution to the problem of finding an absolute date for past volcanic eruptions. Furthermore, it provided a basis for two hypotheses as to how a potential volcanic signature might be derived in tree rings. The next step of the research was to develop a new dendrochemical procedure to apply in this context. This would have the main aim of identifying some form of chemical signature which could be positively connected with a volcanic eruption of known date in a modern, absolutely dated, sequence of tree rings. Once established, the further aim was to explore the possibilities for using this modern analogue to approach the dating of key undated volcanic eruptions of pre-history.

Over the course of the next chapter, the methodology developed for the sampling and analysis of tree rings for the practical part of this project is described. Underlying every step of the design protocol and the selection of analytical techniques is the firm foundation provided by this dendrochemical review.

2.9 Tables

TABLE 2.1: Elements which *increase* at the heartwood sapwood boundary in particular species.

Element	Species	Reference
N	*Quercus robur* L.	[118]
Na	*Lindera erythrocarpa* Makino	[108]
	Liriodendron tulipifera L.	[108]
	Magnolia obovata Thunb.	[108]
	Phellodendron amurense Rupr.	[108]
	Stewartia pseudo-camellia Maxim.	[108]
	Zelkova serrata Makino	[108]
Mg	*Acanthopanax sciadophylloides* Fr. & Sav.	[108]
	Betula grossa Siebold & Zucc.	[108]
	Castanea crenata Siebold & Zucc.	[108]
	Liriodendron tulipifera L.	[108]
	Magnolia obovata Thunb.	[108]
	Quercus robur L.	[118]
	Zelkova serrata Makino	[108]

P	*Larix decidua* Mill.	[103]
	Quercus robur L.	[118]
Cl	*Liriodendron tulipifera* L.	[108]
	Magnolia obovata Thunb.	[108]
	Quercus serrata Thunb.	[108]
	Zelkova serrata Makino	[108]
K	*Liriodendron tulipifera* L.	[108]
	Magnolia obovata Thunb.	[108]
	Phellodendron amurense Rupr.	[108]
	Quercus robur L.	[118]
	Quercus serrata Thunb.	[108]
	Zelkova serrata Makino	[108]
Ca	*Acanthopanax sciadophylloides* Fr. & Sav.	[108]
	Betula grossa Siebold & Zucc.	[108]
	Lindera erythrocarpa Makino	[108]
	Liriodendron tulipifera L.	[108]
	Magnolia obovata Thunb.	[108]
	Quercus robur L.	[118]
	Quercus serrata Thunb.	[108]
	Zelkova serrata Makino	[108]
Mn	*Acanthopanax sciadophylloides* Fr. & Sav.	[108]
	Castanea crenata Siebold & Zucc.	[108]
	Lindera erythrocarpa Makino	[108]
	Liriodendron tulipifera L.	[108]
	Quercus serrata Thunb.	[108]
	Zelkova serrata Makino	[108]
Cu	*Acanthopanax sciadophylloides* Fr. & Sav.	[108]
	Pinus densiflora	[156]
Zn	*Acanthopanax sciadophylloides* Fr. & Sav.	[108]
	Lindera erythrocarpa Makino	[108]
Br	*Quercus serrata* Thunb.	[108]
	Zelkova serrata Makino	[108]
Rb	*Magnolia obovata* Thunb.	[108]
	Phellodendron amurense Rupr.	[108]
	Quercus serrata Thunb.	[108]
	Stewartia pseudo-camellia Maxim.	[108]
	Zelkova serrata Makino	[108]
Sr	*Acanthopanax sciadophylloides* Fr. & Sav.	[108]
	Magnolia obovata Thunb.	[108]
	Quercus serrata Thunb.	[108]
	Zelkova serrata Makino	[108]
Sb	*Cryptomeria japonica* D. Don	[156]
Cs	*Hovenia dulcis* Thunb.	[108]
	Phellodendron amurense Rupr.	[108]
	Zelkova serrata Makino	[108]
Ba	*Lindera erythrocarpa* Makino	[108]
	Zelkova serrata Makino	[108]

TABLE 2.2: Elements which *decrease* at the heartwood sapwood boundary in particular species

Element	Species	Reference

Na	*Magnolia obovata* Thunb.	[108]
	Sorbus commixta Hedl.	[108]
Mg	*Kalopanax pictus* Nakai	[108]
	Quercus alba L.	[159]
Al	*Stewartia pseudo-camellia* Maxim.	[108]
P	*Quercus alba* L.	[159]
K	*Acanthopanax sciadophylloides* Fr. & Sav.	[108]
	Sorbus alnifolia K. Koch	[108]
Ca	*Hovenia dulcis* Thunb.	[108]
	Quercus alba L.	[159]
	Sorbus alnifolia K. Koch	[108]
	Stewartia pseudo-camellia Maxim.	[108]
Mn	*Kalopanax pictus* Nakai	[108]
	Quercus alba L.	[159]
	Stewartia pseudo-camellia Maxim.	[108]
Fe	*Kalopanax pictus* Nakai	[108]
Cu	*Kalopanax pictus* Nakai	[108]
Zn	*Stewartia pseudo-camellia* Maxim.	[108]
Sr	*Sorbus alnifolia* K. Koch	[108]
Ba	*Sorbus alnifolia* K. Koch	[108]
	Stewartia pseudo-camellia Maxim.	[108]
La	*Castanea crenata* Siebold & Zucc.	[108]
Sm	*Castanea crenata* Siebold & Zucc.	[108]

TABLE 2.3: Elements which *peak* at the heartwood sapwood boundary in particular species

Element	Species	Reference
Na	*Betula grossa* Siebold & Zucc.	[108]
	Castanea crenata Siebold & Zucc.	[108]
	Hovenia dulcis Thunb.	[108]
	Quercus serrata Thunb.	[108]
Mg	*Hovenia dulcis* Thunb.	[108]
	Liriodendron tulipifera L.	[99]
	Phellodendron amurense Rupr.	[108]
Al	*Pinus koraiensis* Siebold & Zucc.	[28]
P	*Liriodendron tulipifera* L.	[99]
S	*Quercus alba* L.	[159]
Cl	*Hovenia dulcis* Thunb.	[108]
	Kalopanax pictus Nakai	[108]
	Lindera erythrocarpa Makino	[108]
	Phellodendron amurense Rupr.	[108]
K	*Betula grossa* Siebold & Zucc.	[108]
	Castanea crenata Siebold & Zucc.	[108]
	Hovenia dulcis Thunb.	[108]
	Liriodendron tulipifera L.	[99]
Ca	*Picea rubens* Sarg.	[141]
Mn	*Hovenia dulcis* Thunb.	[108]
	Phellodendron amurense Rupr.	[108]
	Pinus koraiensis Siebold & Zucc.	[28]
	Quercus mongolica Fisch. & Turcz.	[28]
Cu	*Picea mariana* (Mill.) Britton, Sterns & Poggenb.	[132]

Zn	*Magnolia obovata* Thunb.	[108]
Br	*Castanea crenata* Siebold & Zucc.	[108]
	Hovenia dulcis Thunb.	[108]
	Lindera erythrocarpa Makino	[108]
Rb	*Castanea crenata* Siebold & Zucc.	[108]
	Hovenia dulcis Thunb.	[108]
Sr	*Phellodendron amurense* Rupr.	[108]
Cd	*Abies religiosa* H.B.K. & Cham	[164]
	Pinus koraiensis Siebold & Zucc.	[28]
Ba	*Phellodendron amurense* Rupr.	[108]
Sm	*Hovenia dulcis* Thunb.	[108]
	Phellodendron amurense Rupr.	[108]
Pb	*Abies religiosa* H.B.K. & Cham	[164]
	Pinus sylvestris L.	[162]

TABLE 2.4: Elements displaying a decrease in concentrations from pith to bark in various tree species.

Element	Species	Reference
Na	*Pinus elliottii*	[110]
Mg	*Acer pseudoplatanus* L.	[163]
	Acer saccharum Marsh.	[46, 167]
	Larix decidua Mill.	[103]
	Picea mariana (Mill.) Britton, Sterns & Poggenb.	[6]
	Picea rubens Sarg.	[141, 171]
	Pinus elliottii	[110]
	Pinus sylvestris L.	[153]
	Quercus robur L.	[118]
Al	*Picea mariana* (Mill.) Britton, Sterns & Poggenb.	[6]
	Picea rubens Sarg.	[17]
	Pinus sylvestris L.	[153]
	Tsuga canadensis (L.) Carrière	[17]
K	*Abies sp.* L.	[56]
	Acer saccharum Marsh.	[167]
	Pinus elliottii	[110]
	Pinus sylvestris L.	[162]
Ca	*Acer pseudoplatanus* L.	[163]
	Acer saccharum Marsh.	[46, 167]
	Larix decidua Mill.	[103]
	Picea abies (L.) H.Karst.	[122]
	Picea mariana (Mill.) Britton, Sterns & Poggenb.	[6]
	Picea rubens Sarg.	[141, 171]
	Pinus elliottii	[110]
	Pinus sylvestris L.	[153]
	Quercus alba L.	[159]
Ti	*Picea mariana* (Mill.) Britton, Sterns & Poggenb.	[6]
Cr	*Picea abies* (L.) H.Karst.	[122]
	Picea mariana (Mill.) Britton, Sterns & Poggenb.	[6]
Mn	*Acer pseudoplatanus* L.	[163]
	Acer saccharum Marsh.	[46]
	Betula sp.	[138]
	Diospros kaki Thunb.	[4]
	Larix decidua Mill.	[103]

	Picea abies (L.) H.Karst.	[122]
	Picea mariana (Mill.) Britton, Sterns & Poggenb.	[6]
	Pinus elliottii	[110]
	Pinus sp.	[138]
	Pinus sylvestris L.	[153]
Co	*Picea abies* (L.) H.Karst.	[122]
Cu	*Abies balsamea* (L.) Mill.	[129]
	Picea mariana (Mill.) Britton, Sterns & Poggenb.	[6]
	Populus marilandica Bosc & Poir.	[91]
Zn	*Abies balsamea* (L.) Mill.	[129]
	Betula sp.	[138]
	Fagus sylvatica L.	[57]
	Picea abies (L.) H.Karst.	[122]
	Picea mariana (Mill.) Britton, Sterns & Poggenb.	[6]
	Pinus sylvestris L.	[153]
	Populus balsamifera L.	[91]
Rb	*Betula sp.*	[138]
	Cryptomeria japonica D. Don	[4]
	Pinus elliottii	[110]
Sr	*Acer pseudoplatanus* L.	[163]
	Picea abies (L.) H.Karst.	[122]
	Picea mariana (Mill.) Britton, Sterns & Poggenb.	[6]
Cd	*Fagus sylvatica* L.	[57]
	Pinus sylvestris L.	[153]
Cs	*Pinus elliottii*	[110]
Ba	*Picea abies* (L.) H.Karst.	[122]
La	*Picea rubens* Sarg.	[17]
	Tsuga canadensis (L.) Carrière	[17]
Pb	*Abies balsamea* (L.) Mill.	[129]
	Fagus sylvatica L.	[57]
	Pinus sylvestris L.	[153]

TABLE 2.5: Elements displaying an increase in concentration from pith to bark in various tree species.

Element	Species	Reference
Mg	*Pinus sylvestris* L.	[162]
	Quercus alba L.	[89, 159]
	Quercus robur L.	[162]
Al	*Alnus rubra* Bong.	[171]
	Picea mariana (Mill.) Britton, Sterns & Poggenb.	[6]
	Pinus elliottii	[110]
P	*Larix decidua* Mill.	[103]
	Pinus contorta x banksiana (Loud.) Lamb.	[84]
	Pinus sylvestris L.	[153]
	Quercus alba L.	[89, 159]
S	*Fagus sylvatica* L.	[153]
	Pinus sylvestris L.	[153]
K	*Larix decidua* Mill.	[103]
	Picea mariana (Mill.) Britton, Sterns & Poggenb.	[6]
	Pinus sylvestris L.	[153]
	Quercus alba L.	[89]
	Quercus robur L.	[162]

Ca	*Pinus sylvestris* L.	[162]
Cr	*Larix decidua* Mill.	[113]
Mn	*Abies religiosa* H.B.K. & Cham	[164]
	Quercus alba L.	[89, 159]
	Quercus robur L.	[162]
Fe	*Alnus rubra* Bong.	[171]
	Picea mariana (Mill.) Britton, Sterns & Poggenb.	[6]
	Pinus sylvestris L.	[153]
Ni	*Abies religiosa* H.B.K. & Cham	[164]
	Larix decidua Mill.	[113]
	Pinus sylvestris L.	[153]
Cu	*Abies religiosa* H.B.K. & Cham	[164]
	Acer pseudoplatanus L.	[163]
	Acer rubrum L.	[166]
	Larix decidua Mill.	[113]
	Picea rubens Sarg.	[171]
	Pinus sylvestris L.	[153]
Zn	*Abies religiosa* H.B.K. & Cham	[164]
Cd	*Larix decidua* Mill.	[113]
Ba	*Picea mariana* (Mill.) Britton, Sterns & Poggenb.	[6]
Pb	*Larix decidua* Mill.	[113]

TABLE 2.6: Elements displaying relatively stable concentrations from pith to bark in various tree species.

Element	Species	Reference
Mg	*Picea mariana* (Mill.) Britton, Sterns & Poggenb.	[132]
Al	*Larix decidua* Mill.	[103]
Si	*Liriodendron tulipifera* L.	[99]
S	*Quercus alba* L.	[159]
Ca	*Picea mariana* (Mill.) Britton, Sterns & Poggenb.	[132]
Fe	*Picea mariana* (Mill.) Britton, Sterns & Poggenb.	[132]
Ni	*Picea mariana* (Mill.) Britton, Sterns & Poggenb.	[132]
Co	*Pinus elliottii*	[110]
Cu	*Cryptomeria japonica* D. Don	[4]
	Larix decidua Mill.	[103]
	Picea mariana (Mill.) Britton, Sterns & Poggenb.	[132]
Zn	*Larix decidua* Mill.	[103]
	Picea mariana (Mill.) Britton, Sterns & Poggenb.	[132]
	Pinus elliottii	[110]
Br	*Pinus elliottii*	[110]
Rb	*Pinus* sp.	[138]
Sr	*Cryptomeria japonica* D. Don	[4]
Cd	*Quercus robur* L.	[74]
Pb	*Picea mariana* (Mill.) Britton, Sterns & Poggenb.	[132]
	Quercus robur L.	[74]

TABLE 2.7: Examples of approximate concentrations (parts per million) of different elements in various tree species

Element	ppm	Species	Reference
Be	15	*Quercus prinus* L.	[36]
	15	*Quercus rubra* L.	[36]
B	6	*Pinus elliottii*	[110]
Na	25	*Pinus elliottii*	[110]
	50	*Quercus rubra* L.	[36]
	50	*Quercus prinus* L.	[36]
	80	*Pinus strobus* L.	[36]
	100	*Tsuga canadensis* (L.) Carrière	[36]
Mg	332	*Acer pseudoplatanus* L.	[163]
	200	*Acer saccharum* Marsh.	[46]
	100	*Pinus strobus* L.	[46]
	40	*Pinus strobus* L.	[36]
	175	*Quercus prinus* L.	[36]
	199	*Pinus elliottii*	[110]
	50	*Quercus rubra* L.	[36]
Al	6	*Pinus strobus* L.	[36]
	2	*Quercus rubra* L.	[36]
	3	*Quercus prinus* L.	[36]
	4	*Picea abies* (L.) H.Karst.	[122]
	18	*Pinus elliottii*	[110]
Si	11	*Quercus prinus* L.	[36]
P	73	*Acer pseudoplatanus* L.	[163]
	90	*Pinus elliottii*	[110]
	20	*Pinus strobus* L.	[46]
S	50	*Pinus strobus* L.	[46]
	100	*Acer saccharum* Marsh.	[46]
K	350	*Pinus strobus* L.	[46]
	900	*Acer saccharum* Marsh.	[46]
	473	*Pinus elliottii*	[110]
Ca	300	*Pinus strobus* L.	[46]
	600	*Picea abies* (L.) H.Karst.	[122]
	1500	*Acer saccharum* Marsh.	[46]
	30	*Pinus strobus* L.	[36]
	622	*Pinus elliottii*	[110]
	1432	*Acer pseudoplatanus* L.	[163]
	100	*Quercus rubra* L.	[36]
Cr	1	*Picea abies* (L.) H.Karst.	[122]
	8	*Larix decidua* Mill.	[113]
Mn	150	*Abies religiosa* H.B.K. & Cham	[164]
	20	*Pinus strobus* L.	[46]
	150	*Quercus rubra* L.	[36]
	50	*Acer saccharum* Marsh.	[46]
	40	*Pinus strobus* L.	[36]
	150	*Quercus prinus* L.	[36]
	27	*Acer pseudoplatanus* L.	[163]
	107	*Pinus elliottii*	[110]
Fe	2	*Picea abies* (L.) H.Karst.	[122]
	18	*Pinus elliottii*	[110]
	90	*Acer saccharum* Marsh.	[46]
	350	*Pinus strobus* L.	[46]
Ni	3	*Abies religiosa* H.B.K. & Cham	[164]
	6	*Larix decidua* Mill.	[113]
Cu	10	*Larix decidua* Mill.	[113]
	2	*Dacrycarpus dacrydioides*	[147]

	1	*Pinus ponderosa*	[114]
	15	*Acer pseudoplatanus* L.	[163]
	3	*Abies religiosa* H.B.K. & Cham	[164]
Zn	7	*Pinus strobus* L.	[46]
	5	*Acer saccharum* Marsh.	[46]
	7	*Pinus ponderosa*	[114]
	3	*Dacrycarpus dacrydioides*	[147]
	4	*Pinus elliottii*	[110]
	21	*Acer pseudoplatanus* L.	[163]
	300	*Abies religiosa* H.B.K. & Cham	[164]
Br	1	*Pinus elliottii*	[110]
Rb	2	*Pinus elliottii*	[110]
Sr	6	*Quercus prinus* L.	[36]
	4	*Picea abies* (L.) H.Karst.	[122]
	8	*Acer pseudoplatanus* L.	[163]
	5	*Quercus rubra* L.	[36]
	2	*Pinus strobus* L.	[36]
	5	*Pinus ponderosa*	[114]
Cd	2	*Larix decidua* Mill.	[113]
	1	*Acer pseudoplatanus* L.	[163]
Ba	1	*Pinus ponderosa*	[114]
	2	*Pinus strobus* L.	[36]
Pb	2	*Quercus* sp.	[152]
	5	*Carya* sp.	[124]
	150	*Acer pseudoplatanus* L.	[158]
	160	*Ulmus procera* L.	[158]
	100	*Platanus orientalis* L	[158]
	150	*Fraxinus excelsior* L.	[158]
	100	*Aesculus hippocastanum* L.	[158]
	6	*Larix decidua* Mill.	[113]
	2	*Acer pseudoplatanus* L.	[163]
	2	*Abies religiosa* H.B.K. & Cham	[164]
	90	*Quercus robur* L.	[158]

TABLE 2.8: Specific anatomical associations with concentrations of various elements. Complied from [6, 7, 14, 17, 19, 65, 90, 103, 110, 132, 150, 159, 168].

	B	Na	Mg	Al	Si	P	S	Cl	K	Ca	Cr	Mn	Fe	Ni	Cu	Zn	As	Cs	Ba	Pb
Epithelial Cell Walls									Y	Y		Y	Y		Y	Y				
Late Wood									Y	Y		Y			Y					
Middle Lamella										Y	Y	Y	Y		Y	Y				
Early Wood			Y							Y	Y					Y				
Pit Membranes					Y					Y										
Cell Corners										Y	Y	Y			Y				Y	
Ray							Y			Y	Y	Y	Y		Y	Y				
Resin Canal															Y					
Patches		Y		Y			Y	Y	Y				Y		Y	Y			Y	
Tracheid Walls			Y	Y					Y	Y		Y	Y	Y	Y	Y				Y
Bark			Y		Y					Y			Y			Y		Y		
Torus										Y	Y	Y	Y		Y	Y				Y
Pith	Y		Y																	

TABLE 2.9: A list of the different analytical techniques used in dendrochemical studies

Analytical Technique	References
Anodic stripping voltammetry	[91]
Atomic Absorption Spectroscopy	[28, 37, 41, 46, 49, 57, 74, 113, 118, 124, 129, 140, 147, 151, 152, 153, 158]
Direct Coupled Plasma Atomic Emission Spectroscopy	[141]
Electron microprobe analyzer-scanning microscope	[159]
Energy Dispersive X-Ray Fluorescence	[17, 93, 92, 138]
Flame Atomic Emission Spectroscopy	[103]
Fourier transformation infrared spectroscopy	[64]
Gamma-spectrometry	[12]
High-field nuclear magnetic resonance spectroscopy	[64]
High-performance liquid chromatology	[64]
Inductively Coupled Plasma Atomic Emission Spectroscopy	[6, 36, 46, 89, 153]
Inductively Coupled Plasma Mass Spectrometry Laser Ablation	[65, 66, 122, 167, 166, 165]
Inductively Coupled Plasma Mass Spectrometry Solutions	[58, 114, 162, 163, 164, 161, 165]
Inductively Coupled Plasma Optical Emission Spectroscopy	[5, 17, 18, 54]
Inductively Coupled Plasma Scan	[28]
Neutron Activation Analysis	[4, 6, 49, 55, 63, 88, 108, 110, 121, 171]
Particle Induced X-Ray Emission	[4, 50, 90, 84, 99]
Secondary Ion Mass Spectrometry	[7, 6, 14, 19, 97]
Synchrotron Radiation Microbeam X-Ray Fluorescence	[14, 150]
Transmission Electron Microscopy - Energy Dispersive X-Ray Analysis	[132, 168]
Tri-Carb liquid Scintillation Spectrometry	[86]
X-Ray Fluorescence	[81, 154]

TABLE 2.10: A table to show which elements have been detected by which analytical techniques from a sample of the literature

Elements	Technique	References
Cu, Zn	Anodic stripping voltammetry	[91]
Mg, P, S, K, Ca, Cr, Mn, Fe, Ni, Cu, Zn, As, Cd, Pb	Atomic Absorption Spectroscopy	[28, 37, 41, 46, 49, 57, 74, 113, 118, 124, 129, 140, 147, 151, 152, 153, 158]
Mg, P, S, Cl, Ca, Mn	Electron microprobe analyzer-scanning microscope	[159]
Mg, Al, K, Ca, Mn, Fe, Cu, Zn, Rb, Pb	Energy Dispersive X-Ray Fluorescence	[17, 81, 93, 92, 138, 154]
Mg, Al, P, Cu, Zn	Flame Atomic Emission Spectroscopy	[103]
Na, Mg, Al, Si, P, S, K, Ca, Ti, Cr, Mn, Fe, Ni, Cu, Zn, Sr, Ba, Pb	ICP Atomic Emission Spectroscopy	[6, 36, 46, 89, 141, 153]

Mg, Al, K, Ca, Cr, Mn, Fe, Ni, Co, Cu, Zn, As, Sr, Cd, Sn, Ba, Hg, Pb	ICP Mass Spectrometry Laser Ablation	[122, 165, 167, 166]
Na, Mg, Al, P, K, Ca, Mn, Ni, Cu, Zn, Rb, Sr, Cd, Ba, La, Ce, Nd, Sm, Gd, Tm, Lu, Pb	ICP Mass Spectrometry Solutions	[58, 114, 162, 163, 164, 161, 165]
Mg, Al, Ca, Fe, Zn, Cd, La, Pb	ICP Optical Emission Spectroscopy	[5, 17, 18, 54, 153]
Al, Ti, V, Cr, Mn, Fe, Ni, Cu, Zn, Cd	ICP Scan	[28]
B, Na, Mg, Al, P, S, Cl, K, Ca, Mn, Fe, Cu, Zn, Br, Rb, Sr, Mo, Cs, Ba, La, Sm	Neutron Activation Analysis	[4, 6, 55, 88, 108, 110, 134, 171]
Mg, Al, Si, P, S, Cl, K, Ca, Sc, Mn, Fe, Ni, Cu, As, Cu, Rb, Sr, Ba	Particle Induced X-Ray Emission	[4, 50, 90, 84, 99]
Na, K, Ca, Cr, Mn, Fe, Ni, Cu, Zn, As, Cd, Pb	Secondary Ion Mass Spectrometry	[6, 7, 14, 19, 97]
K, Ca, Mn, Fe, Cu, Zn, Ba	Synchrotron Radiation Microbeam X-Ray Fluorescence	[14, 150]
Na, Mg, Al, S, Cl, K, Ca, Cr, Mn, Fe, Ni, Cu, Zn, Pb	TEM - Energy Dispersive X-Ray Analysis	[132, 168]
Pb	Tri-Carb liquid Scintillation Spectrometry	[86]

TABLE 2.11: A sample of the various species of tree used in dendrochemical studies

Species	Reference
Abies balsamea (L.) Mill.	[88, 129]
Abies religiosa H.B.K. & Cham	[164]
Abies sp. L.	[56]
Acanthopanax sciadophylloides Fr. & Sav.	[108]
Acer pseudoplatanus L.	[158, 162, 163, 165]
Acer rubrum L.	[166]
Acer saccharum Marsh.	[96, 69, 161, 167, 170]
Aesculus hippocastanum L.	[86, 158]
Alnus glulinosa L.	[86]
Alnus rubra Bong.	[92, 93, 171]
Betula grossa Siebold & Zucc.	[108]
Betula pubescens L.	[86]
Betula sp.	[138]
Castanea crenata Siebold & Zucc.	[108]
Cryptomeria japonica D. Don	[4, 156]
Carya sp.	[124]
Celtis australis	[157]
Dacrycarpus dacrydioides	[147]
Diospros kaki Thunb.	[4]
Fagus sylvatica L.	[57, 86, 118, 153]
Fraxinus excelsior L.	[158]
Hovenia dulcis Thunb.	[108]
Juniperus virginiana L.	[54, 55]
Kalopanax pictus Nakai	[108]
Larix decidua Mill.	[103, 113]
Lindera erythrocarpa Makino	[108]

Liriodendron tulipifera L.	[99, 108]
Magnolia obovata Thunb.	[108]
Phellodendron amurense Rupr.	[108]
Picea abies (L.) H.Karst.	[14, 64, 122, 146, 150]
Picea mariana (Mill.) Britton, Sterns & Poggenb.	[6, 132]
Picea rubens Sarg.	[17, 37, 141, 168, 171]
Picea sitchensis (Bong.) Carr.	[14]
Picea sp.	[90, 138]
Pinus banksiana Lamb.	[7]
Pinus contorta x banksiana (Loud.) Lamb.	[84]
Pinus densiflora	[156]
Pinus echinata Mill.	[5]
Pinus elliottii	[110]
Pinus koraiensis Siebold & Zucc.	[28]
Pinus ponderosa	[114]
Pinus sp.	[138]
Pinus strobus L.	[49, 97]
Pinus sylvestris L.	[12, 162, 63, 151, 153]
Pinus taeda L.	[154]
Platanus orientalis L	[158]
Populus balsamifera L.	[91]
Populus marilandica Bosc & Poir.	[91]
Pseudotsuga menziesii (Mirb.) Franco	[58]
Quercus alba L.	[89, 159]
Quercus mongolica Fisch. & Turcz.	[28]
Quercus nigra L.	[50]
Quercus prinus L.	[36]
Quercus robur L.	[41, 74, 118, 158, 162]
Quercus rubra L.	[36]
Quercus serrata Thunb.	[108]
Quercus sp.	[58, 152]
Sequoiadendron giganteum (Lindl.) Buchholz	[58]
Sorbus alnifolia K. Koch	[108]
Sorbus commixta Hedl.	[108]
Stewartia pseudo-camellia Maxim.	[108]
Taxodium distichum (L.) Rich.	[81, 154]
Tilia europaea L.	[86]
Tsuga canadensis (L.) Carrière	[17, 36]
Ulmus glabra Huds.	[86]
Ulmus procera L.	[158]
Zelkova serrata Makino	[108]

TABLE 2.12: Elements which have been found to be highly mobile in certain species of trees

Element	Species	Reference
Mg	*Picea abies* (L.) H.Karst.	[122]
P	*Pinus elliottii*	[110]
	Pinus sylvestris L.	[63]
Zn	*Fagus sylvatica* L.	[57]
Mo	*Pinus sylvestris* L.	[63]
Cd	*Fagus sylvatica* L.	[57]
Pb	*Fagus sylvatica* L.	[57]
	Picea rubens Sarg.	[37]

TABLE 2.13: Elements which have been found to be of moderate mobility in certain species of trees

Element	Species	Reference
Al	*Larix decidua* Mill.	[103]
	Pinus koraiensis Siebold & Zucc.	[28]
P	*Liriodendron tulipifera* L.	[99]
K	*Liriodendron tulipifera* L.	[99]
	Pinus contorta x banksiana (Loud.) Lamb.	[84]
	Pinus elliottii	[110]
Ca	*Picea abies* (L.) H.Karst.	[122]
Cr	*Picea abies* (L.) H.Karst.	[122]
Mn	*Abies balsamea* (L.) Mill.	[88]
	Picea abies (L.) H.Karst.	[122]
	Pinus koraiensis Siebold & Zucc.	[28]
	Quercus mongolica Fisch. & Turcz.	[28]
Co	*Picea abies* (L.) H.Karst.	[122]
Cu	*Acer pseudoplatanus* L.	[163]
	Larix decidua Mill.	[103]
Zn	*Abies balsamea* (L.) Mill.	[88]
	Picea abies (L.) H.Karst.	[122]
	Pinus sylvestris L.	[151]
	Taxodium distichum (L.) Rich.	[81]
Br	*Pinus elliottii*	[110]
Sr	*Picea abies* (L.) H.Karst.	[122]
Cd	*Pinus koraiensis* Siebold & Zucc.	[28]
	Pinus sylvestris L.	[151]
	Quercus robur L.	[41]
Ba	*Picea abies* (L.) H.Karst.	[122]
Pb	*Pinus sylvestris* L.	[162]
	Quercus robur L.	[162]
	Quercus robur L.	[41]

TABLE 2.14: Elements which have been found to be of low mobility in certain species of trees

Element	Species	Reference
B	*Pinus sylvestris* L.	[63]
N	*Pinus sylvestris* L.	[63]
Mg	*Pinus elliottii*	[110]
	Pinus sylvestris L.	[63]
Al	*Picea abies* (L.) H.Karst.	[122]
	Pinus echinata Mill.	[5]
	Pinus sylvestris L.	[63]
	Quercus mongolica Fisch. & Turcz.	[28]
S	*Pinus sylvestris* L.	[63]
K	*Pinus sylvestris* L.	[63]
Ca	*Pinus echinata* Mill.	[5]
	Pinus elliottii	[110]
	Pinus sylvestris L.	[63]
Ti	*Pinus koraiensis* Siebold & Zucc.	[28]
V	*Pinus koraiensis* Siebold & Zucc.	[28]

Cr	*Pinus koraiensis* Siebold & Zucc.	[28]
Mn	*Picea rubens* Sarg.	[171]
	Pinus echinata Mill.	[5]
	Pinus elliottii	[110]
	Pinus sylvestris L.	[63]
Fe	*Picea abies* (L.) H.Karst.	[122]
	Pinus koraiensis Siebold & Zucc.	[28]
Ni	*Pinus koraiensis* Siebold & Zucc.	[28]
Cu	*Pinus echinata* Mill.	[5]
	Pinus koraiensis Siebold & Zucc.	[28]
	Pinus sylvestris L.	[151]
Zn	*Larix decidua* Mill.	[103]
	Picea rubens Sarg.	[171]
	Pinus echinata Mill.	[5]
	Pinus sylvestris L.	[63]
	Quercus mongolica Fisch. & Turcz.	[28]
As	*Pinus koraiensis* Siebold & Zucc.	[28]
Rb	*Pinus sylvestris* L.	[63]
Cd	*Acer pseudoplatanus* L.	[163]
	Picea abies (L.) H.Karst.	[122]
	Pinus echinata Mill.	[5]
	Quercus mongolica Fisch. & Turcz.	[28]
Pb	*Picea abies* (L.) H.Karst.	[122]
	Pinus koraiensis Siebold & Zucc.	[28]
	Pinus sylvestris L.	[151]
	Taxodium distichum (L.) Rich.	[81]

TABLE 2.15: Elements which have been successfully linked to deposition from specific environmental pollution sources, from a range of different studies

Element	Reference
Na	[36, 58, 97]
Mg	[17, 18, 36, 118, 141]
Al	[17, 36, 84, 171]
Si	[84]
P	[84]
S	[46, 55, 84, 153]
Cl	[84]
K	[118]
Ca	[17, 18, 36, 58, 118, 141]
Mn	[36, 49, 58, 84, 164, 171]
Fe	[5, 84, 171]
Ni	[84, 164]
Cu	[58, 84, 129, 147, 151, 164]
Zn	[58, 81, 84, 129, 147, 163, 164]
As	[84]
Rb	[58]
Sr	[36, 89]
Mo	[55]
Cd	[41, 54, 147, 163, 164, 165]
La	[58]
Ce	[58]

CHAPTER 2: DENDROCHEMISTRY

Nd	[58]	
Sm	[58]	
Gd	[58]	
Tm	[58]	
Lu	[58]	
Pb	[41, 46, 49, 54, 81, 92, 124, 129, 147, 151, 154, 157, 158, 162, 163, 164, 167, 165]	

TABLE 2.16: The response of various element concentrations in different tree species to a documented increase in environmental acidity

Response	Element	Species	Reference
Decrease	Mg	*Fagus sylvatica* L.	[118]
		Quercus rubra L.	[36]
	P	*Acer pseudoplatanus* L.	[163]
	K	*Fagus sylvatica* L.	[118]
		Quercus robur L.	[118]
	Ca	*Quercus rubra* L.	[36]
		Quercus robur L.	[118]
		Fagus sylvatica L.	[118]
	Sr	*Quercus prinus* L.	[36]
	Mo	*Juniperus virginiana* L.	[55]
Increase	Na	*Tsuga canadensis* (L.) Carrière	[36]
	Mg	*Picea rubens* Sarg.	[17]
		Acer pseudoplatanus L.	[163]
		Picea abies (L.) H.Karst.	[122]
		Picea rubens Sarg.	[141]
	Al	*Quercus rubra* L.	[36]
		Pinus contorta x banksiana (Loud.) Lamb.	[84]
		Picea rubens Sarg.	[17]
		Pinus sylvestris L.	[63]
	Si	*Pinus contorta x banksiana* (Loud.) Lamb.	[84]
	S	*Pinus contorta x banksiana* (Loud.) Lamb.	[84]
		Acer saccharum Marsh.	[46]
	Cl	*Pinus contorta x banksiana* (Loud.) Lamb.	[84]
	K	*Pinus sylvestris* L.	[63]
	Ca	*Acer pseudoplatanus* L.	[163]
		Picea rubens Sarg.	[17]
		Picea rubens Sarg.	[141]
	Cr	*Acer saccharum* Marsh.	[69]
	Mn	*Abies balsamea* (L.) Mill.	[88]
		Acer pseudoplatanus L.	[163]
		Acer saccharum Marsh.	[69]
		Quercus rubra L.	[36]
	Fe	*Pinus contorta x banksiana* (Loud.) Lamb.	[84]
	Ni	*Acer saccharum* Marsh.	[69]
		Pinus contorta x banksiana (Loud.) Lamb.	[84]
	Co	*Acer saccharum* Marsh.	[69]
	Cu	*Pinus sylvestris* L.	[151]
		Acer pseudoplatanus L.	[163]
		Pinus contorta x banksiana (Loud.) Lamb.	[84]

	Zn	*Acer saccharum* Marsh.	[69]
		Pinus contorta x banksiana (Loud.) Lamb.	[84]
		Abies balsamea (L.) Mill.	[88]
	As	*Pinus contorta x banksiana* (Loud.) Lamb.	[84]
		Acer saccharum Marsh.	[69]
	Rb	*Pinus sylvestris* L.	[63]
	Sr	*Acer pseudoplatanus* L.	[163]
	Cd	*Abies religiosa* H.B.K. & Cham	[164]
		Juniperus virginiana L.	[54]
	Pb	*Acer pseudoplatanus* L.	[163]
		Pinus sylvestris L.	[151]
		Abies religiosa H.B.K. & Cham	[164]
		Juniperus virginiana L.	[54]
None	P	*Pinus contorta x banksiana* (Loud.) Lamb.	[84]
	K	*Pinus contorta x banksiana* (Loud.) Lamb.	[84]
	Ca	*Pinus contorta x banksiana* (Loud.) Lamb.	[84]
	Mn	*Pinus sylvestris* L.	[63]
	Zn	*Pinus sylvestris* L.	[63]
	Rb	*Pinus contorta x banksiana* (Loud.) Lamb.	[84]
	Sr	*Pinus contorta x banksiana* (Loud.) Lamb.	[84]

Chapter 3

Development of Methodology

A major part of this project has been focused on the development of new methodologies for the improved sampling and analysis of various wood types. This chapter introduces the various stages and outcomes of this developmental process. It will begin by discussing the optimised field sampling strategy for obtaining the best wood samples for chemical analysis. Next, the three analytical techniques selected for the investigation of the tree ring chemistry will be introduced. Finally, the specific areas of methodological design required and implemented to successfully couple the specific analytical techniques with the analysis of various wood types will be discussed.

3.1 Field sampling

A critical part of any experimental design is the selection and collection of the optimum samples to meet the specifications of the research objective. In this context, issues for consideration include the species of tree, the condition of the tree and the nature of the growth environment. Also the way in which these factors might relate to the environmental / atmospheric impact of any particular volcanic eruption. The type of samples taken should also be considered to ensure that there is sufficient material for replication, and to represent the whole tree and preferably several others at any one site. There is also the issue of storing the samples between collection and analysis. The following, optimum procedure for sampling living trees was developed based primarily on practical experimentation as part of the overall project and an amalgamation of methodologies from other research groups.

1: Select a suitable species for analysis (see section 2.6.2 for further details).

2: Select trees which are free from any damage or decay. Either of these can obviously have huge consequences for the chemistry rendering them useless for environmental monitoring of atmospheric composition [160].

3: If possible, select a tree growing in an environment where dominant uptake of atmospheric chemistry is likely to be via bark and leaves rather than through the soil. Poor, thin soils, based on slow eroding bedrock in a dry environment are likely to encourage more accurate recording of environmental deposition patterns from direct deposition. The site should be relatively flat in order to avoid the presence of tension or compression wood formed where trees grow on sloping ground or where there are strong, dominant wind directions. The structure of this wood is affected, and some researchers [122] have found this to be accompanied by a change in chemistry which would hinder the tracing of external deposition. The site should also be topographically situated so as to benefit from a wide catchment of deposition.

4: Several trees from each site should be sampled from several locations within each tree so that the chemistry can be investigated at various heights and orientations in the trunk. Standardise all sampling procedures, taking samples from the same heights and orientations from similar aged/sized trees.

5: If sampling trees close to a specific volcano (or other environmental pollutant source), use a similar control site to see if any natural radial tendencies in element concentration can be observed in the same species in a similar environment, not exposed to the same degree of atmospheric deposition.

6: Whilst sampling, record other environmental data such as soil depth, the nature of the underlying bedrock and the vegetation cover. Where possible take abiotic samples for further analysis and comparison with the chemistry of the wood. Find out dominant wind directions and precipitation regimes. Also note all sampling equipment contacting the samples (for example the corer or chain-saw) in order to fully consider possible sources of contamination later on. For soil sampling, the standard procedure developed for this project was to, where possible, clear a representative profile of the substrate down to bed rock. An annotated diagram was then made in the field describing any horizons in terms of colour (Munsell), stoniness, percentage organic material and any interpretation as to the mineralogy of the rock and soil. Also note any evidence suggesting which horizon provided the dominant source of uptake for the sampled tree (i.e. the presence of tree roots). Soil samples were taken by digging up to five pits around the base of the tree and sampling from each apparent horizon in each pit. The samples from individual horizons from each pit were then amalgamated and mixed to produce an averaged sample for that horizon.

7: Immediately after sampling, seal samples in aerated plastic containers to minimise contamination and limit fungal growth by encouraging air circulation. If the mode of sampling is coring, samples can be stored effectively in plastic drinking straws which have previously been aerated with a pin. If the sampling was carried out with a chain saw, larger samples (cookies) can be stored in aerated plastic bags. Ideally samples should be frozen immediately after sampling to limit the migration of sap, however this has not been proven necessary, and is likely to be impractical in most field situations. The most important part of sample storage between collection and final analysis is keeping the samples free from the growth of mould. In the timber industry this is achieved by storing wood in a cold, dry environment for a number of years, however, again, this is not likely to be a practical option for many dendrochemical studies. It was found that the best results were achieved by drying both cores and cookies at room temperature in a low moisture environment. Samples were propped in breathable racks so that air was able to circulate round the whole sample. To avoid radial cracking cookies were subsampled and quarters removed to provide more flexibility for contraction. Some cracking did occur however, but this was not sufficient to be problematic to the the dendrochemical or dendrochronological techniques employed to study the samples.

The reality of the project in terms of time and financial constraints, meant that external sources were relied upon for the collection of the majority of samples. Opportunities to put the optimised sampling strategy into practice were few and far between (for the main example see chapter 5). Many of the samples obtained came from existing dendrochronologies from around the globe (see list of contributors in the acknowledgements section). In this case specific details of the original sampling sites (for example topography and substrate of the growth environment) were often not known, either because they were not required for the study from which the samples came or, more often because the wood was not sampled in situ, but came from some sort of preservation environment. The implications of studying samples from an environment of preservation are illustrated in chapter 6.1, samples may be contaminated by this environment, or commonly, by fungal growth and or procedures used to prepare the samples for dendrochronological counting such as sanding and chalking. All samples obtained for the project were recorded in a database along with associated information on the sample source and any results from the analysis of the sample. If dendrochemical techniques could be demonstrated to successfully detect evidence of global or localised volcanism in modern wood, the condition and availability of dendrochronological samples for prehistory would then become an issue and the problems arising with dendrochronological samples previously discussed may become especially relevant. This matter is further discussed in chapter 7.

3.2 Analytical techniques

Once samples were obtained, a series of further sampling and processing steps were taken before analysis via one or more of the selected analytical techniques. This section will introduce these techniques and then go on to describe the associated developed methodologies.

Inductively Coupled Plasma Mass Spectrometry (ICP-MS) combines ultra low detection limits of less than 0.05ng per ml for the majority of elements, with rapid analysis of around three to five minutes per sample for simultaneous, multi-element determination of the majority of elements in the periodic table. A plasma is a highly ionised form of matter at a very high temperature, composed of electrons and atomic nuclei. An inductively coupled plasma is, in this case, generated by electrodeless discharge (to prevent contamination) in a gas at atmospheric pressure. It is maintained by energy coupled to it from a radio frequency generator. Argon (Ar) gas is used because it is inert - i.e. will not form stable compounds with analytes and it has a high ionisation energy. The plasma produced is very stable, generating a constant amount of energy and thus providing a constant stream of ions. An *Argon plasma* can ionise practically every element. Samples are aspirated into the plasma via the stream of Ar gas where they are ionised. They emerge as undissociated molecular fragments, unvolatised particles, atoms and ions. These then pass through into a conventional mass spectrometer, where, as every element has a specific mass to charge ratio, the ions are detected and counted. The typical, quadrupole spectrometer can make 0-300 scans of the entire spectrum in one hundred microseconds, which is relatively high speed. However the latest technology is the magnetic sector system which has much higher resolution with simultaneous transmission of a limited (frequently nine or ten) suite of masses. The only downside of this new system is that it has a slower data acquisition rate.

Samples can be introduced to this system in two main ways, either via the conventional induction of samples in solution, or in the form of a suspension generated via laser ablation of a solid sample to remove a small amount of microparticulate material which subsequently becomes entrained in the argon gas flow and injected straight into the plasma (LA-ICP-MS).

Inductively Coupled Plasma Atomic Emission Spectrometry (ICP-AES) is one of the most widely used techniques of elemental analysis. As with ICP-MS, an inductively coupled plasma excitation source is used, however in this case the temperature in the part of the flame used is lower. This heat is not sufficient to ionise the atoms, instead the excited electrons drop back to their normal state. As they do so, electromagnetic emission of wavelengths specific to each individual element are emitted. Results are produced by a system which counts the emissions. The method can scan most elements but not with such a range or down to the trace amounts detected by ICP-MS, in terms of precision though, the detection achieved is actually more precise.

ICP-MS analysis was carried out on a quadrupole, VG Thermo Elemental PlasmaQuad ICP-MS. For LA-ICP-MS this was used in conjunction with a Cetax LSX-100 laser (Nd:YAG pulsed with Q-switch) operating at 266 nm. These were driven by PQVision version 4.1.2 and Cetac laser system version 1.20, with a high resolution CCD camera system for observation of the sample during analysis. ICP-AES was carried out on a Perkin Elmer Optima 3000 ICP with AS90 Auto Sampler.

LA-ICP-MS was the main technique initially selected for the purposes of this project. It was selected as a new, state of the art technique, the potential of which for the analysis of tree rings had not been fully tested. The prospect of analysing at annual and even sub-annual resolution for a wide suite of elements including the rare earth elements, and to do so rapidly with minimal sample preparation or destruction, made it appear the ideal analytical technique for the purpose. Development and application, however, revealed certain problems with the use of LA-ICP-MS for the analysis of wood, and so conventional ICP-MS was employed as a complementary and / or alternative analytical method. ICP-AES was selected primarily with a view to detecting sulphur, which cannot be accurately detected by ICP-MS as there are lots of interferences with oxygen. In addition to analysing for S, ICP-AES was also used to produce independently generated data for comparison with ICP-MS data sets for various elements easily detected by both methodologies.

3.2.1 Methodological design requirements: Laser Ablation ICP-MS

The main design issues for the use of LA-ICP-MS for the analysis of tree rings were sample preparation, data processing procedures and calibration. With these in place, improving the reproducibility of data became the main focus for development. As only very small quantities of original sample (which can be directly sub-sampled from an existing solid sample) are required for LA-ICP-MS, going from the original sample to analysis is a relatively straightforward process. However, in terms of sample preparation, certain steps were required to produce clean sub-samples, mounted and marked in order to facilitate navigation of the sample in the laser ablation chamber.

Whilst basic steps of the data processing procedure were similar to those used to correct most raw analytical data, a certain amount of fine tuning was required in order to produce the best possible calculated data for wood. This included the addition of an internal standardisation (normalisation) calculation and various correction and calibration methodologies. In order to facilitate the data processing procedure it was also necessary to design a series of spreadsheets which could be used to speed up the calculation procedure and to re-calculate various data sets using different calibration methodologies.

The main area for methodological design in the whole LA-ICP-MS process was calibration. Given the relative newness of the technique, and the fact that it was designed for application to rock samples, calibration standards are not readily available for most organic matrices. Ideally, wood calibration standards should have the same matrix as the wood tissue being sampled, however such standards are not commercially available. Standards which are available, however can be used to produce semi-quantitative data, though this requires the addition of several steps of calculations to the data processing procedure. Three different calibration methodologies were developed during the course of the project. Two methods for semi-quantitative calibration were developed using existing glass standards and an attempt to produce quantitative results was made with the production of a specifically tailored calibration pellet. The final issue to be investigated (rather than designed), was a failure to replicate data sets for the same tree rings. All these various steps in methodological development are discussed in section 3.3.

3.2.2 Methodological design requirements: Solutions ICP-MS

The main area for methodological development for the successful use of solutions ICP-MS for the analysis of wood, lay in terms of dissecting wood samples and bringing them satisfactorily into solution. Problems to address included the fact that such sample preparation procedures are generally time consuming and that the process brings in numerous opportunities for contamination. There was also the issue of developing a successful procedure for the total dissolution of samples. This is complex as different fractions within the same sample may digest in different ways and at different rates; this can mean that more volatile elements are lost as part of the digestion process. Digestion methodologies must therefore be fine tuned to ensure all elements of interest are retained. Where dissolution is not complete, samples cannot be analysed without filtration, this introduces a new opportunity for contamination and results in a loss of sample which is difficult to quantify.

Another aspect of the fine tuning is finding the right dilution factor for the samples: this must dilute the sample matrix sufficiently to be run through the instrument without over diluting the sample which may result in the total loss, or loss of precision of certain lower concentration analytes. The methodologies developed to overcome many of these issues are discussed in section 3.4, in terms of both the dissection and the dissolution of the wood samples.

3.2.3 Methodological design requirements: ICP-AES

ICP-AES is a well established analytical technique which has been tried and tested over the years on many different sample types. Therefore the only areas for methodological development were, as with solutions ICP-MS, the dissection and dissolution of various types of wood sample. The methodological development is therefore also considered in section 3.4.

3.3 Methodological Design: Laser Ablation ICP-MS

3.3.1 Sample preparation

Whilst sample preparation for LA-ICP-MS samples is minimal, it became clear throughout the development of this methodology and during analysis, that a number of preparation steps could be introduced to improve the quality of collected data sets and to facilitate rapid, effective sampling in the sample chamber. As a result the following procedure was developed. Samples were cut to fit within the dimensions of the laser sampling chamber. If a long sequence of years was desirable, samples were cut into thin strips and mounted alongside one another in order of years (see figure 3.1 for an example). Samples were examined under times twenty magnification and the sampling surface was cleaned via removal of the old surface with a sterile steel blade or, where necessary (for harder woods) a steel sledge microtome in order to produce the flattest, smoothest possible sampling surface (for further explanation of the cleaning procedure see section 3.4.1 and figure 3.11 for illustrations). A steel pin was used to mark each ring or sub-section of a ring in order to facilitate navigation to correct sampling sites in the sample chamber. A sketch was made of the sample prior to analysis, marking the start and finish years, the direction of analysis and the position of the sample within the sample chamber. A specific set-up was designed for the laser ablation system delimiting the analytical conditions under which all samples subsequently were run. As this differs little from other potential set-up scenarios for an LA-ICP-MS system, it is only included as Appendix D.

3.3.2 Data processing

Raw data from LA-ICP-MS analysis are expressed as the ion counts detected by the mass spectrometer. In order to convert these data into meaningful numbers in concentrations or to be expressed as ratios, a series of calculations had to be applied. The first two steps in the developed data processing procedure were basic and would apply to most raw data. These were blank subtraction and the removal of values less than the lower limit of detection. In laser ablation, the procedural blanks are measured when replicate analyses are made of each element in the gas passing through the sample chamber and a mean value (excluding statistical anomalies) is calculated. The average counts measured in gas blanks run throughout a day's analysis were then subtracted from the counts for the corresponding sample data. This corrects for background levels of various elements and polyatomic species. A lower limit of detection (LoD, in counts) was then calculated as three times the standard deviation of the replicate gas blanks. All values less than this were then discarded as below the lower limit of detection - i.e. falling within the overall level of variation and therefore not significant (those remaining have above 99% confidence (3 σ) that there is a true value). This step cancels out low levels of counts from samples which may represent background noise rather than a true concentration.

As the complex cell structures of various species of wood do not provide an homogeneously structured plane for ablation, differing quantities of material were sampled by each standard laser shot. This meant it was impossible to determine whether low or high counts for a given sample were simply related to true fluctuations in elemental concentrations, or just to the total amount of material sampled. The next step in the data correction procedure was therefore to correct for this effect. An internal standard was required, that is, an element with a constant concentration throughout each sample, to which other elements could subsequently be related. Carbon was selected for this purpose. Carbon accounts for approximately 50% of wood [1]. It was found through several trial analyses that the counts for carbon, measured by the ^{13}C isotope, remained

---
[1]This varies slightly from tree to tree, and from species to species, for example *Quercus* sp. have on average 48.95% and *Pinus* sp. around 49.91%, [98]

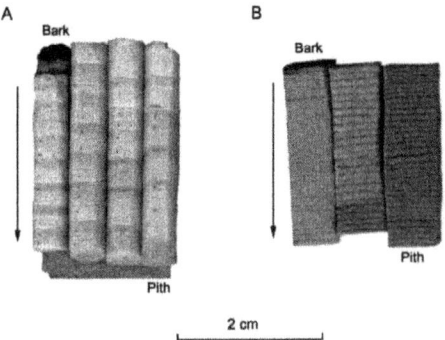

FIGURE 3.1: A: Sample core, B: Sub-section of a radial sample. Both are mounted for laser ablation sampling. Arrows note direction of sampling from bark to pith

the most constant through various analyses of different wood types when the laser was fully absorbed. The ^{12}C isotope was rejected as it is too abundant and was observed to cause signal saturation (i.e. too much signal for the instrument to detect) ^{13}C forms only 1% of total carbon, and so provides a more reliable signal. Assuming that concentrations of ^{13}C are sufficiently constant, then each analysis of a given sample should contain equal counts. The proportional difference between one sample and another is the ratio by which all other elements must be corrected. ^{13}C counts for the first run (visually checked to have ablated a typical area) were divided by the counts for ^{13}C in the subsequent runs. This produced a factor which was then applied to correct the counts for all other elements in that run, i.e. all elements in all samples were normalised to ^{13}C. So, if two areas of one wood sample were ablated, the first area giving 100,000 ^{13}C counts and 500 counts for ^{88}Sr, and the other (lower density area) giving 50,000 ^{13}C counts and 250 counts ^{88}Sr, then assuming the ^{13}C concentrations should be the same, a ratio of 2:1 or 0.5 can then be applied to the ^{88}Sr for the second run to reveal that the two samples in fact contained the same proportion of ^{88}Sr.

The next two data processing steps, as with the first, followed a more standard procedure for fine tuning LA-ICP-MS data. These were to correct for isotopic abundance and ionisation efficiency. Different isotopes have different abundances, if an element has more than one isotope the number of counts for one mass will not represent the total amount of that element in the sample. This was corrected for by bringing each isotope up to the equivalent of 100% abundance, by taking the number of counts multiplied by one hundred and dividing by the percentage isotopic abundance. The Ionisation efficiency is the ease with which an element will form positively charged ions in the plasma and is based on their first ionisation potential. The data were corrected so that it was as though 100% of all atoms put into the plasma were ionised and consequently detected. This was achieved by taking the number of counts multiplied by one hundred, and dividing by the percentage ionisation efficiency. The percentage ionisation efficiency can be measured using a mass spectrometer or by studying the emission spectra of atoms. The first ionisation potential is a measurement of the amount of energy required to remove one mole of electrons from one mole of gaseous atoms. Successive ionisation energies for the same element measure the energy required to remove a second, third, fourth, fifth electron and so on. For this exercise pre-measured values of ionisation efficiency, produced by the NERC ICP Lab at Kingston-Upon-Thames, were used.

The final steps of the data processing procedure were to convert the corrected counts into concentrations in the wood. This was attempted via three different calibration strategies (see section 3.3.3). All stages of data processing were automated on specially designed Excel spreadsheets. The spreadsheet design evolved with the methodology so several versions exist allowing data to be processed via different calibration methodologies. At the beginning of each processed data sheet there is a properties page which includes the spreadsheet version. This system effectively facilitated and greatly speeded up data processing, allowing the same data sets to be processed using different calibration methods. A database was also designed to manage the thousands of data sets generated and to promote easy navigation between data sets.

3.3.3 Calibration

The first calibration methodology attempted involved the construction and application of a semi-quantitative calibration curve. The curve was produced by ablation of NIST SRM 610 and 612 glass wafers [117]. These are silicate glass reference materials produced by the National Institute of Standards and Technology (NIST), proven to be homogenous and spiked with up to 61 trace elements at nominal concentrations of $500\mu g$ (610) and $50\mu g$ (612). A new curve was constructed for each run of samples as, owing to day to day fluctuations in instrumental mass response, the strength of the signal differed slightly on each day of analysis. NIST discs were ablated prior to each run of samples, and a spread of the detected counts for selected elements across the mass range were used to calculate a best-fit curve using regression equations. The second order calibration curve generated could then be used to find the response in ppm in relation to the mass of any specific element. If the subsequently run samples were physically very similar to NIST discs (which were designed for the calibration of rock samples), then the conversion of the detected counts for the samples would have been almost fully calibrated, and the concentrations in ppm would have attained a good level of accuracy. However, as wood ablates very differently to glass, the concentrations generated were not exact, but semi-quantitative data.

One further step was added to this procedure in an attempt to correct for the differences in the ablation of wood versus the ablation of a glass. It was noted that the apparent carbon concentrations produced by this process were higher than they should have been based on the assumption that wood is 50% carbon. By this reasoning the maximum carbon concentration detected in any sample must be 500 000 ppm (1% = 10,000 ppm). Anything greater than this must be due to the different ablation properties of wood and glass. The difference between this figure (50% or a close average for any given species e.g. *Pinus* sp. 49.91% [98])
and the calculated carbon content at this point is the degree of exaggeration. This is compensated for by finding the number by which to multiply the calculated values for carbon content to reduce it to 50%. For example, if ^{13}C = 3,542,934 then: 500,000 / 3,542,934 = factor y (in this case 0.141). By applying this factor to all the calculated data a more accurate idea of the true concentrations can be gained. This principle was also employed to check the quality of the NIST glass curve by using calcium (a known constant in the discs) instead of carbon as an internal standard. NIST discs were also used in the general sampling procedure (regardless of the calibration method) to monitor instrumental drift after each batch of ten samples.

When data calibrated by this method were studied one problem which became apparent was that it was not possible to join continuing sequences of data together from one day to the next without a pronounced step appearing at the join. This was because of day-to-day differences in the sensitivity of the instrument. An example of this is given in figure 3.2 where Ca and V data are plotted from a continuous sequence of the same *Pinus contorta* sample, collected on two separate days of analysis.

Whilst this effect is obvious, impacting most elements in the same way, and can be accounted for in terms of data interpretation, it was far from desirable. The advantages of fast analysis of a hundred or more tree rings in a day would be offset if chronologies of several hundred years could not be strung together in a seamless way. Other problems noted at this point were difficulties in replicating data sets generated from the same wood sample, and the fact, that out of the wide suite of elements scanned for, only a handful were reliably detected. It was decided that the first and second of these problems might be combated by the design of a simplified calibration process.

This new calibration methodology produced final data, not in concentrations, but in ratios to ^{13}C. Data were processed as previously, blanks were subtracted, values below the lower limit of detection were removed and the data were corrected for percentage abundance and the degree of ionisation. The only difference was that data were no longer normalised to ^{13}C (the main cause of the 'stepped' effect observed in the previous data sets). All elements in each sample

CHAPTER 3: DEVELOPMENT OF METHODOLOGY

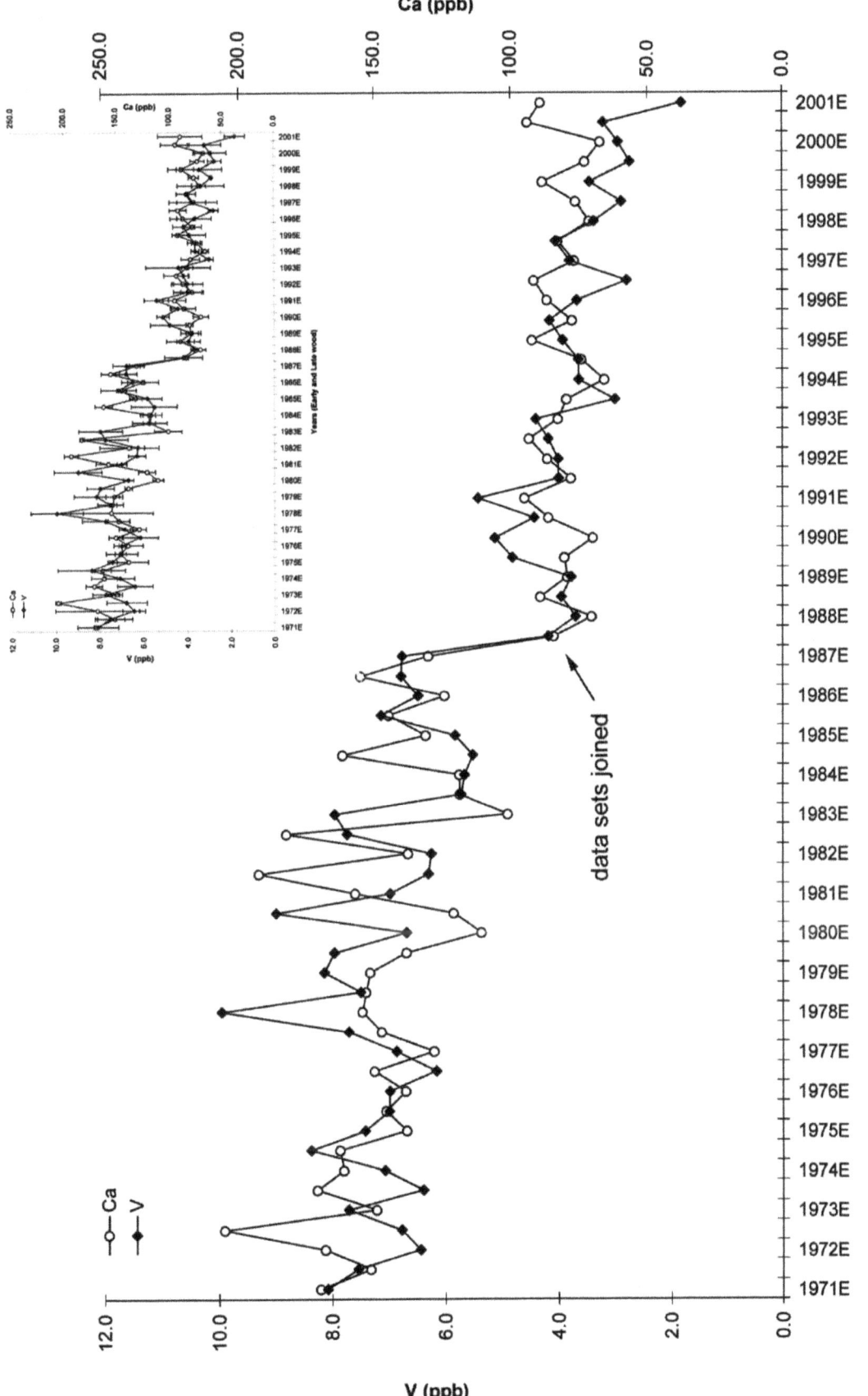

FIGURE 3.2: Data for Ca and Vanadium (V) from sample series 48 and 51. Data sets are joined in 1987. There are two values per year, representing high resolution ablations of the early and late wood in order to look at differences in seasonal increments. Inset shows data with error bars

on a particular day were ratioed to the ^{13}C in the same sample and the NIST discs were ratioed to the ^{13}C which was retained from the gas blank. (NIST data for ^{13}C were not blank subtracted, as there is no carbon in NIST, the ^{13}C in the air was required as a value for calibration). This was achieved by dividing the counts for each element by the counts for ^{13}C. In order to compensate for differences in the daily detection of the instrument, *master* values were produced for the NIST discs. These were the averages of fully processed NIST runs for all elements taken on one, arbitrarily selected day when instrument performance levels were optimum. All subsequently produced data sets were then calibrated back to this set of master NIST standard values, to improve the linkage between continuous data sets from different days of analysis. A calibration factor was produced by dividing the overall mean from the master NIST data with the overall mean of the NIST ablations for a day.

Figure 3.3 shows a comparison of the data set for Ca (as in figure 3.2) calibrated to a NIST disc to give semi-quantitative concentrations in parts per billion, and presented via the new methodology in ratios to ^{13}C. This is repeated for the Mn data from the same sample. These results show that the ratio method facilitates the successful linking together of data sets. It can also be used to correct for differences between the overall ratios of replicate data sets processed on different days, thus facilitating a more direct comparison in order to determine the degree of replication. Processing the same data sets by the two different calibration methods also served as a check confirming that both methods re-calculate the data in a reliable way.

The problem with producing data in ratios however is that it is not directly comparable with data sets generated by alternative methods of chemical analysis. There was no way of improving the precision of the data collection via this method. Therefore, as neither methodology was perfect, further possibilities for refinement were explored. The only foreseeable way to improve both the accuracy and precision of the semi-quantitative method was by developing an alternative calibration standard with similar ablation properties to wood. Only by substitution of the glass curve with a curve generated by a more suitable standard, could fully quantitative calibration be attempted.

Several other research teams had previously attempted to develop calibration strategies with a view to the analysis of tree rings. Early attempts by Watmough *et al.* [165] found linear relationships for Pb, Mn and Mg between dissolved solutions of specific tree rings and laser ablated samples of the same ring [165]. Later papers use individual tree rings as calibration standards [167, 166], this technique was shown to work well with *Acer sp.*(maple) tree rings, however can only work with tree species proven to have homogenous chemistry in a single tree ring. The other most widely attempted method for calibration is the production of specially designed pellets. Alteyrac *et al.* [2] pressed sawdust pellets spiked with a Rhodium (Rh) standard for the calibration of *Quercus* tree rings, however the sawdust was found to be too heterogeneous for accurate calibration. More successful attempts have been made by Prohaska *et al.* and Hoffmann *et al.* [6, 66, 122] who pressed pellets from pure cellulose powder spiked with elemental standards. Whilst such cellulose pellets have still been argued to differ structurally (and chemically) from wood [2], this methodology was selected for further development as it provides a suitable compromise between homogeneity and similarity with wood.

The first part of the process was to work out the maximum amount of liquid standard that could be added to a pellet prior to pressing. This would define the limitation on the number of elements possible in one pellet. Test pellets 1-6 were pressed:

1. 0.5g cellulose at 5 Ton for 5 min

2. 20 μl H$_2$0 and 0.5g cellulose at 5 Ton for 5 min

3. 50 μl H$_2$0 and 0.5g cellulose at 5 Ton for 5 min

4. 80 μl H$_2$0 and 0.5g cellulose at 5 Ton for 5 min

5. 170 μl H$_2$0 and 0.5g cellulose at 5 Ton for 5 min

6. 200 μl H$_2$0 and 0.5g cellulose at 5 Ton for 5 min

7. 200 μl H$_2$0 + 0.5gm cellulose + 5 Ton for 5 min

8. 180 μl H$_2$0 + 0.5gm cellulose + 5 Ton for 5 min

CHAPTER 3: DEVELOPMENT OF METHODOLOGY

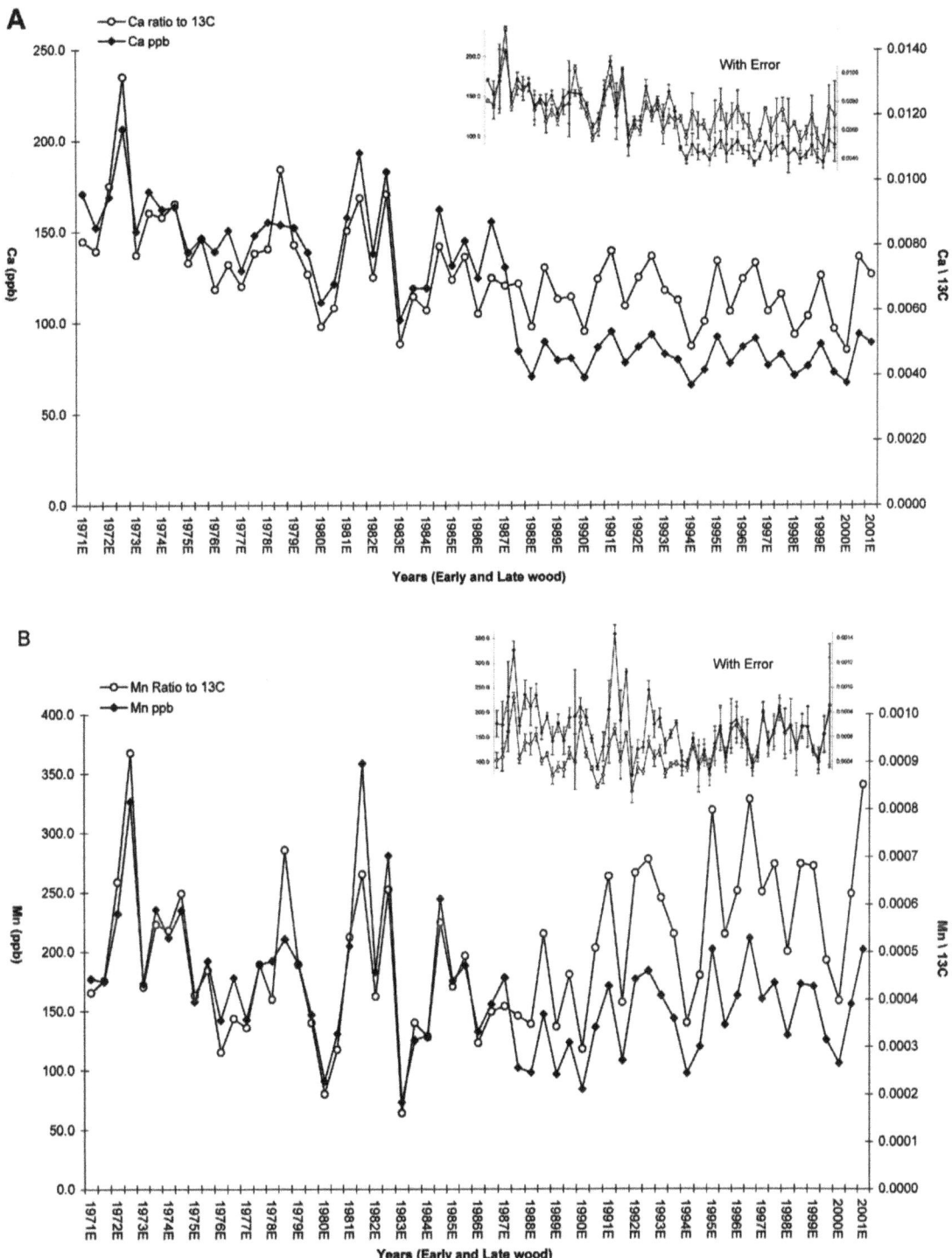

FIGURE 3.3: The same data set processed by both the ratio and the semi-quantitative calibration methodologies. A: Ca data from figure 3.2. B: Mn data from the same data set. Error shown on insets.

9. 150 μl H_2O + 0.5gm cellulose + 5 Ton for 5 min

A start pressure of 5 tons to 0.5g of Aldrich 20μm cellulose was selected based on the methodologies of Prohaska et al. and Hoffman et al. [6, 66, 122]. Pellet six could not hold all the added water and excess droplets were squeezed out. Pellet five however, was pressed successfully. The next step was to check the ablation properties of each of the successfully pressed pellets to see if the water content had any impact on the structure. This step was also intended to observe if the cellulose pellets ablated in a way that was visually similar to wood. The samples were then ablated under the same conditions as the wood samples. Ablation was visually similar to that of *Pinus* early wood. However, after ablation the higher water content pellets displayed cracks around the circumference. A further three pellets (7,8 and 9) were pressed in an attempt to overcome this problem.

These were mixed, and air dried as powder for 15 minutes, then pressed. These pellets ablated successfully without cracking. Five new pellets were made substituting the H_2O for Mn and Ba standard solutions at concentrations of 0ppm, 1ppm, 10ppm, 50ppm and 100ppm. Solutions and powder were mixed, left to dry for 20 minutes then pressed. Pellets ablated with no cracks. The background contribution of the cellulose after blank subtraction and removal of values below the lower limit of detection was found to be suitably low, with over 99% C, the rest being made up of traces of Al, Fe, and Ca. The straight calibration lines produced for the two elements showed that direct calibration was possible via this method. The next step was to produce multielemental calibration pellets. Nine element standards were selected for the pellets; Sr, La, Arsenic (As), Cu, Mn, Ba, Silver (Ag), Nickel (Ni) and Al. 0.1 ml of each of the 10000mg / ml standard solutions were diluted in 1ml of H_2O to produce a stock solution of 50mg / 100ml. Five standard pellets (0 - 100ppm) were made as follows:

- 0 ppm: 0.5g cellulose and 1ml H_2O
- 1 ppm: 0.5g cellulose, 100ml stock and 900ml H_2O
- 10 ppm: 0.5g cellulose, 200ml stock and 800ml H_2O
- 50 ppm: 0.5g cellulose, 500ml stock and 500ml H_2O
- 100 ppm: 0.5g cellulose and 1000ml stock

These were mixed thoroughly with a spatula for 15 minutes then left to dry for a further 5. The calibration lines produced were good for Sr, La, As, Cu, Mn, Ba and Ag. Al and Ni were not successful. Figure 3.4 shows examples of the calibration curves created for Ag, La, Sr, Mn, Ba and Al.

The pellets were then ablated prior to analysis of samples and a semi-quantitative curve was constructed for calibration. Figure 3.5 shows an example of the type of curve generated. Figure 3.6 shows an example of how the curve degraded after two months, with the concentrations of the elements falling at various rates, this underlines the need to produce new calibration pellets for each new batch of analysis. As elements at the same concentration do not show the same intensity of response over the whole mass range in LA-ICP-MS (i.e. there is mass discrimination) the points about which the curve is constructed do not produce a statistically accurate curve. A similar effect was found by Alteyrac et al. [2]. This places the use of a cellulose generated semi-quantitative curve in a very dubious light. The inaccuracy is highlighted by the differences between data for specific elements directly quantified to the pellet, and the same data calibrated to the semi-quantitative curve. An example of this is provided in figure 3.7 where the results for Mn for two wood types are shown. This illustrates the fact that the semi-quantitative data have greater error and also can vary considerably from the directly quantified data.

In order to fully test this calibration method, and to see if it performed better for particular wood types, ten year samples of four different wood types were prepared for solutions and laser ablation analysis. Calibration of the ten years laser ablation data was carried out directly to the cellulose pellet for Sr, La, As, Cu, Mn, Ba and Ag. The same data sets were also calibrated to the semi-quantitative curve generated both from the cellulose pellet and (in the case of Pinus and Quercus) the NIST disc. A range of fifteen elements were selected for comparison between the concentration values produced for each calibration method in relation to the solution derived concentrations (ppm).

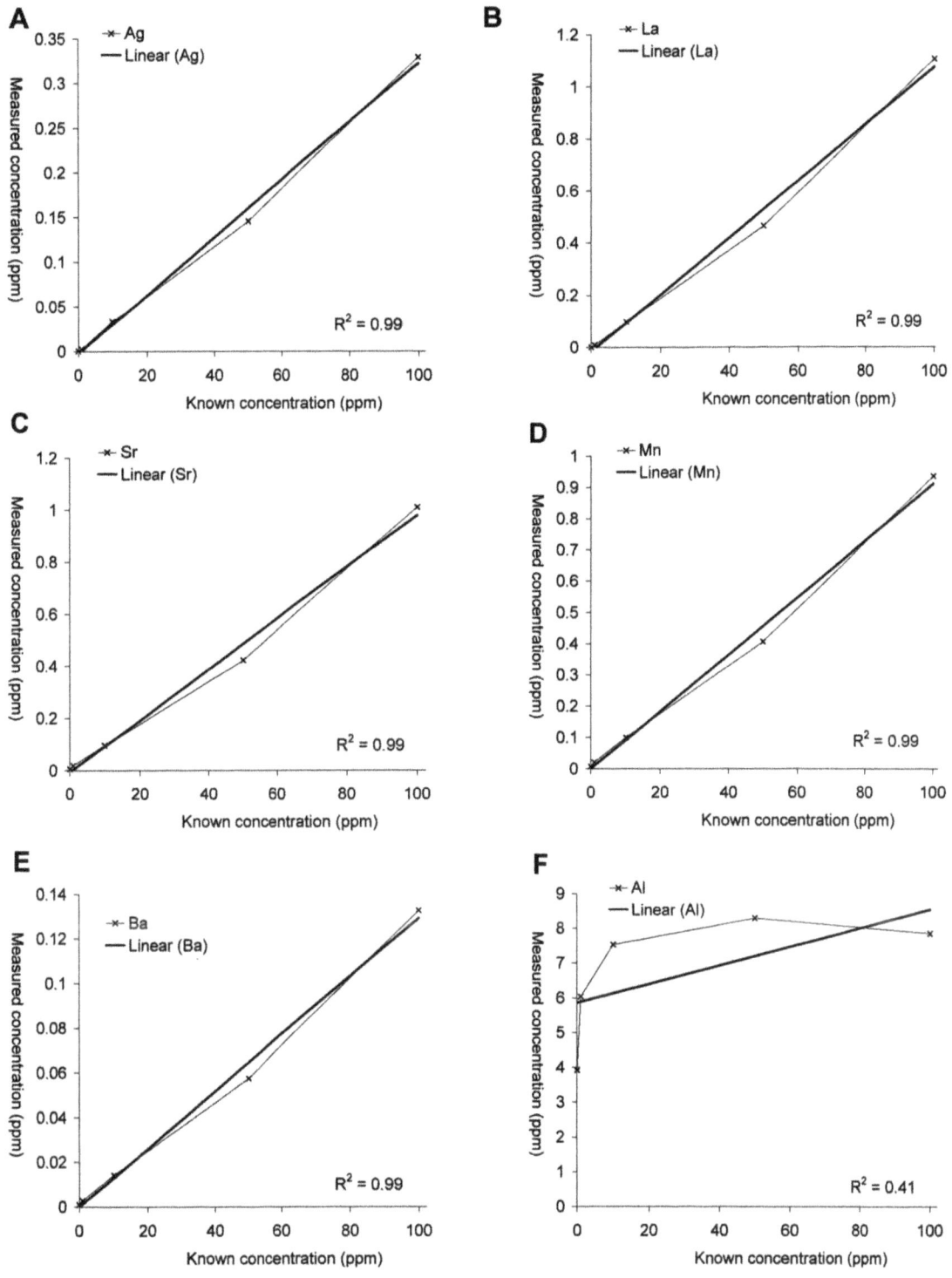

FIGURE 3.4: Calibration lines generated for A: Ag, B: La, C: Sr, D: Mn, E: Ba, and F: Al

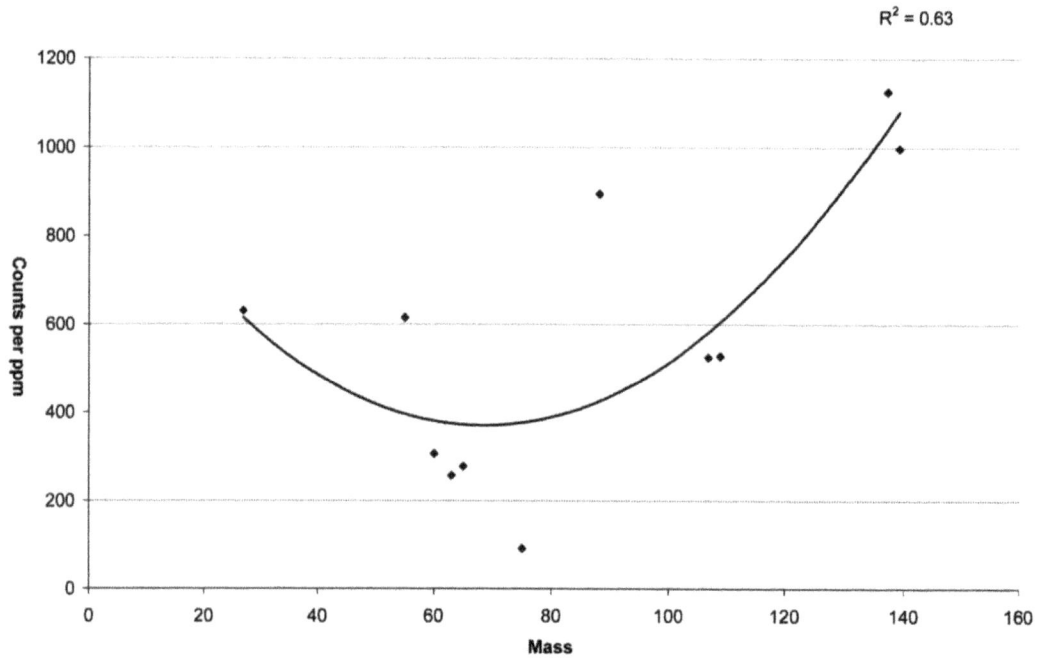

FIGURE 3.5: Semi-quantitative curve

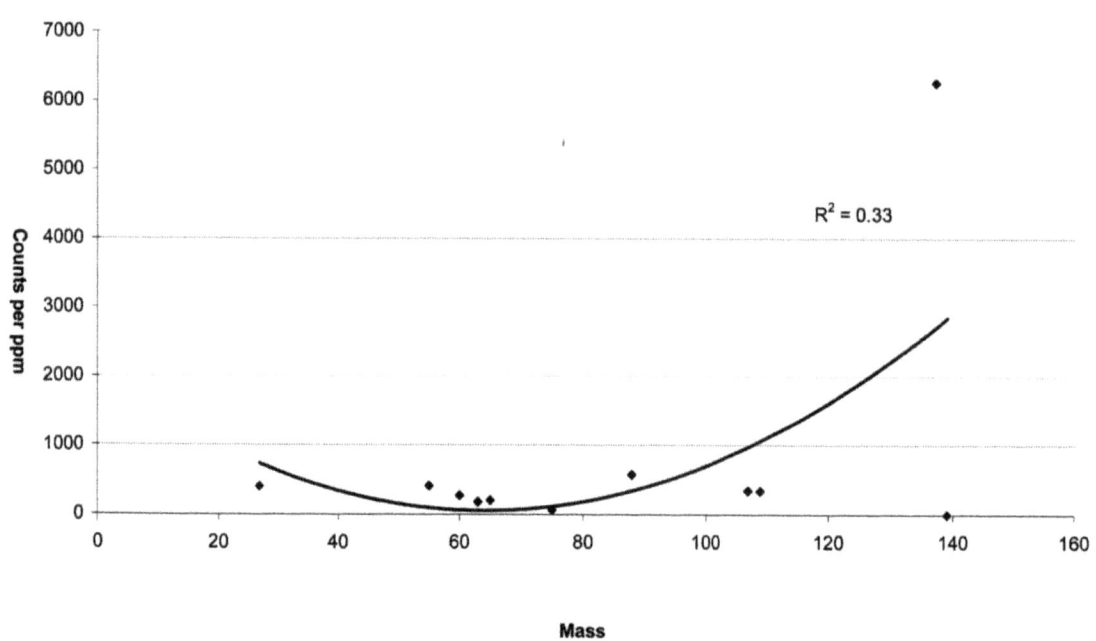

FIGURE 3.6: Semi-quantitative curve

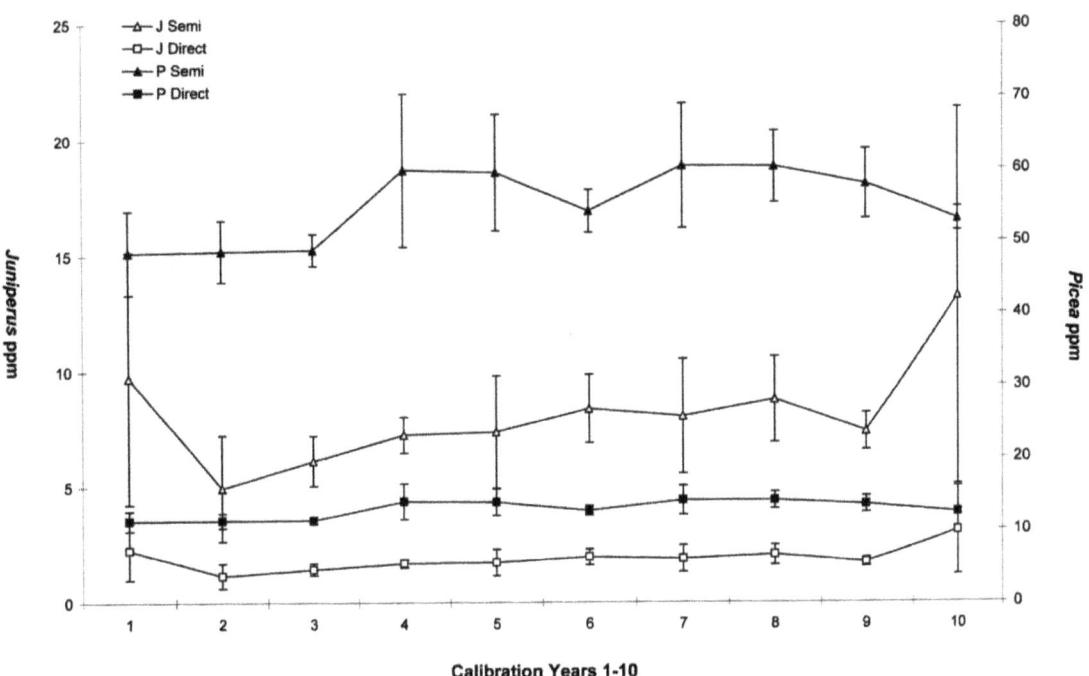

FIGURE 3.7: Direct and semi-quantitatively calibrated data for Mn, for ten years of *Picea* sp. and ten years of *Juniperus* sp. wood samples

These data[2] are presented in tables 3.1 to 3.4 for *Pinus sp.*, *Quercus sp.*, *Juniperus sp.* and *Picea sp.*

Ideally, the different calibration methods would have been expected to produce very similar values to the quantitative concentrations produced for the same tree rings via solutions analysis. As can be seen from the four tables however, this was not the case. Whilst some elements (for example Al and Mn) displayed relatively good agreement for each of the methods for each wood type, many of the other elements (for example Zn and Ca) differed by an order of magnitude or more. Overall the results were highly variable with no really obvious relationships between the data for the different methods, wood types or elements. However certain observations could still be made. Logically speaking the best calibration method should have been via direct quantification to the spiked cellulose pellet, where it was possible to compare between two or three of the calibrated results, this was generally the case. However on a wood to wood basis, *Pinus* displayed an equal number of closer correlated values to the solutions, for both direct and semi-quantitative calibration to the cellulose pellets. *Quercus* on the other hand, showed the closest agreement for the largest number of values between the solutions data and the data which were semi-quantitatively calibrated to the NIST disc. Where comparisons could be made, *Juniperus* and *Picea* were calibrated best via direct calibration to the cellulose. The fact that *Quercus* appeared to calibrate slightly better to the NIST disc than to the cellulose pellet is probably indicative of the fact that, as it is a hardwood, its ablation properties are closer to the glass than the cellulose. The other wood types are all softer and so were more suited to attempted calibration via cellulose.

The data were tested to see if any better correlation between the solution and the calibrated samples of individual years could be observed when the ten year sequences were plotted against one another. An example of one of the most closely correlated data sets is given in figure 3.8. Even for the best calibrated element,

---

[2] All data presented in tables have been corrected for error and rounded to the correct number of significant figures. Error is shown in figures as error bars.

TABLE 3.1: Concentrations in *Pinus sylvestris* via different calibration methods. D-Q and S-Q (direct-quantitative and semi-quantitative). Where 'nd' the element in question was not reliably determined, or was below detection.

Element	Solution	Solution	S-Q Cellulose	S-Q NIST
Al	289	59.9	32.6	nd
Ca	1020	nd	12700	nd
V	0.034	nd	1.7	2.3
Mn	1.1	36	165	103
Ni	0.95	56	nd	nd
Cu	2.7	9.75	30.20	27
Zn	0.00078	50	25	0.5
Rb	4	nd	5.8	26
Sr	0.004	5.2	10.4	106
Y	0.005	nd	0.200	0.16
Zr	0.0008	nd	nd	0.280
Ag	0.09	0.25	0.09	0.1
Cd	0.001	nd	0.35	nd
Sb	0.0005	nd	nd	0.0170
Ba	0.034	nd	0.100	20

TABLE 3.2: Concentrations in *Quercus robur* via different calibration methods. D-Q and S-Q (Direct-quantitative and semi-quantitative). Where 'nd' the element in question was not reliably determined, or was below detection.

Element	Solution	D-Q Cellulose	S-Q Cellulose	S-Q NIST
Al	85	73.3	41.3	124
Ca	130	nd	16200	7
V	0.039	nd	2.8	0.5
Mn	4.2	7.6	31	33
Ni	1.35	73	320	nd
Cu	1.4	11.7	43	nd
Zn	0.002	nd	96	0.9
Rb	2.9	nd	9.3	0.28
Sr	0.001	4.7	7.8	2.9
Y	0.003	nd	0.35	0.05
Zr	0.0003	nd	0.41	0.02
Ag	0.005	nd	nd	0.07
Cd	0.005	nd	0.4	0.05
Sb	0.0017	nd	nd	0.0035
Ba	0.001	nd	0.237	0.02

TABLE 3.3: Concentrations in *Juniperus communis* via different calibration methods. NIST semi-quantitative calibration not available. D-Q and S-Q (Direct-quantitative and semi-quantitative). Where 'nd' the element in question was not reliably determined, or was below detection.

Element	Solution	D-Q Cellulose	S-Q Cellulose
Al	116	47.1	25.1
Ca	434	nd	5780
V	0.4	nd	0.8
Mn	1.2	1.4	6.2
Ni	0.69	nd	nd
Cu	2.3	6.5	18.5
Zn	0.017	nd	25
Rb	3	nd	2.96
Sr	0.0006	4.4	10.6
Y	0.002	nd	nd
Zr	0.0003	nd	nd
Ag	0.002	nd	nd
Cd	0.009	nd	nd
Sb	0.0036	nd	nd
Ba	0.0008	nd	0.019

TABLE 3.4: Concentrations in *Picea glauca* via different calibration methods. NIST semi-quantitative calibration not available. D-Q and S-Q (Direct-quantitative and semi-quantitative). Where 'nd' the element in question was not reliably determined, or was below detection.

Element	Solution	D-Q Cellulose	S-Q Cellulose
Al	1	95	52
Ca	266	nd	18900
V	nd	nd	2.81
Mn	7.2	10.7	55
Ni	0.17	148	636
Cu	0.9	9.9	30.2
Zn	13.3	nd	111
Rb	0.2	nd	6.7
Sr	9.08	8.8	16
Y	0.001	nd	0.65
Zr	0.004	nd	0.98
Ag	0.026	0.23	0.09
Cd	0.007	nd	0.6
Sb	0.002	nd	nd
Ba	9.5	nd	0.29
		nd	

(Mn, via direct quantification to cellulose, and solutions in *Pinus*), little correlation between the yearly concentrations or patterns could be seen. This highlights the failure of LA-ICP-MS not only to produce accurately calibrated concentrations by direct or semi-quantitative calibration to a glass or cellulose standard, but also to produce truly replicable sequences of the same tree rings.

There are a variety of explanations for why the various calibration methods displayed such a variable lack of correlation with the solutions data. The first of these is that the various wood types and the cellulose ablate in very different ways and contain variable concentrations of C. In this case, failure to calibrate could be seen as due to pellet design, and the conclusion would be the need to develop a better calibration pellet, or series of pellets for use with specific wood types. However, in reality it seems another issue, fundamental to all the difficulties experienced with the use of LA-ICP-MS for the analysis of wood may be responsible.

Comparison between solutions of sub-samples and laser ablation of the same sample, is standard procedure in ICP-MS for checking the quality of a calibration method. The assumption is made that the two sample types are of the same elemental composition and any differences noted in the final concentrations must be due to inconsistencies in the calibration method (plus instrumental fluctuations). However, although this might be true for the types of homogenous geological samples for which the technique is often used, it is not necessarily the case for wood. If the individual tree rings sampled were not of a relatively homogeneous composition, then it could be argued that in reality the data for solutions and the data for laser ablation are not directly comparable. A solution of a particular year does not represent exactly the same sample as an ablation of the same year. Sampling with a laser takes only a small ablated fraction of a single tree ring, whereas a solution of the same ring provides an average of a much wider area. If the tree ring does not have homogeneous chemistry the two samples from the same ring could be considerably different. In this case, it may not be so much the problem of unsuitable calibration standards, but that the quality of the calibration may not be tested in this way due to the heterogeneity of the tree rings.

In conclusion, it seems that much further work is required if LA-ICP-MS is ever to be used reliably for the analysis of concentrations in wood. The prospects for creating a wood standard which is chemically and structurally similar to wood are not high, as the material in question is complex and highly variable and its chemistry is poorly understood. The type of data which can be gained currently via the laser ablation of wood are relative data, relating one sample to another. Hypothetically this would be sufficient for the study of changes in the chemistry of sequences of annually dated tree rings, however such data cannot be compared to the work of others, and thus are harder to substantiate. The revelation that the homogeneity of the tree rings may make them impossible to compare with solutions of the same sample is however, of critical importance to any such future developments. The implications spread far further into the potential use of LA-ICP-MS for the analysis of tree rings than merely preventing the production of data in concentrations. The degree of heterogeneity displayed by tree rings could also be directly responsible for how well a sequence of tree rings could be replicated both in terms of concentrations and critically, patterns of relative data. Without the potential to replicate generated data sets the future use of LA-ICP-MS for the analysis of tree ring sequences would be highly questionable. For this reason the final section of methodological development for this analytical procedure will explore this key issue in greater depth.

3.3.4 Tree ring homogeneity and reproducibility

Attempts to replicate various element sequences of tree rings from the same tree, and to replicate samples from within individual tree rings proved difficult. Figure 3.9 shows an example of the typical type of agreement shown by three replicate analyses for the same sequence of tree rings. Although the ablations of the *Pinus contorta* sample were taken adjacent to one another in the same ring for each of the replicated years there is no correlation between them.

This failure to replicate sequences was found for repeated data sets for all elements and for various wood types. One possible explanation was the previously discussed need to develop

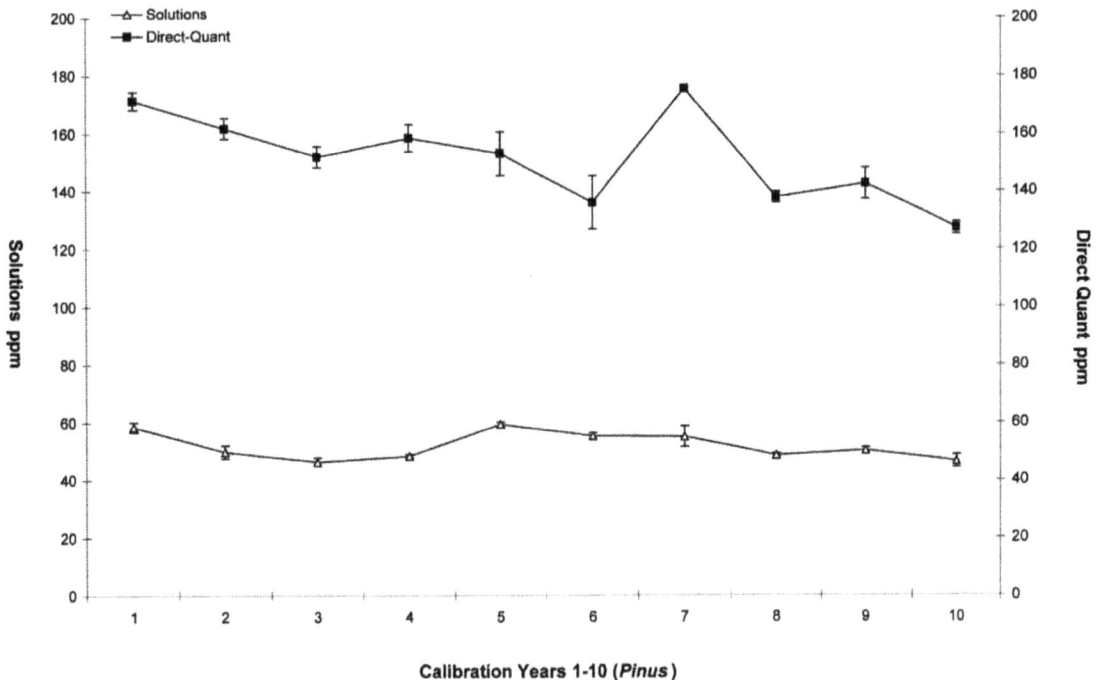

FIGURE 3.8: Solution concentrations versus directly calibrated laser ablation data for the same ten year sequence of *Pinus* sp.

effective means of calculating and calibrating the data sets. However, the most obvious explanation might come from the wood itself.

The structure of wood is extremely complex and intricate and varies greatly from species to species. Could this complexity be reflected in the chemistry of the wood as well? The literature contains various evidence to suggest that the chemistry of individual tree rings can indeed be very complex and vary widely within a very small area of a given sample. If this is the case then perhaps the laser ablation sampling method provides too high a resolution sample of a given tree ring - i.e. perhaps tree rings are too heterogeneous for analysis by LA-ICP-MS.

To investigate this, ten replicate samples were taken from the same area of a single tree ring for two wood samples. The percentage standard deviation of the replicates for each ring was calculated for each element. Ten replicates of an homogenous NIST disc were also made and percentage standard deviations of the ablations were calculated in the same way. These data (see table 3.5) were then compared.

As the NIST disc is known to be homogeneous, the percentage standard deviation of the ten ablations represents the degree of variability introduced by the instrument. By comparing this with the percentage standard deviation for various wood types the heterogeneity of the various tree rings could be better understood.

As would be expected, the percentages are lower for the NIST disc than for the wood for most elements. For these particular tree rings the variation is shown to be low to moderate for most elements. Exceptions are Cu in *Quercus* and Mn in *Pinus* which stand out as being rather variable within a very small area of the tree ring. Two more rings were ablated in the same way from each of the samples. This was to see if different tree rings in the same sample display different degrees of variability. The results are shown in table 3.6. Variability, and even the elements which could be reliably determined for the second tree rings, were very different.

From further mini studies such as these, it was concluded that the chemistry of individual tree rings from the same sample of wood could be

CHAPTER 3: DEVELOPMENT OF METHODOLOGY

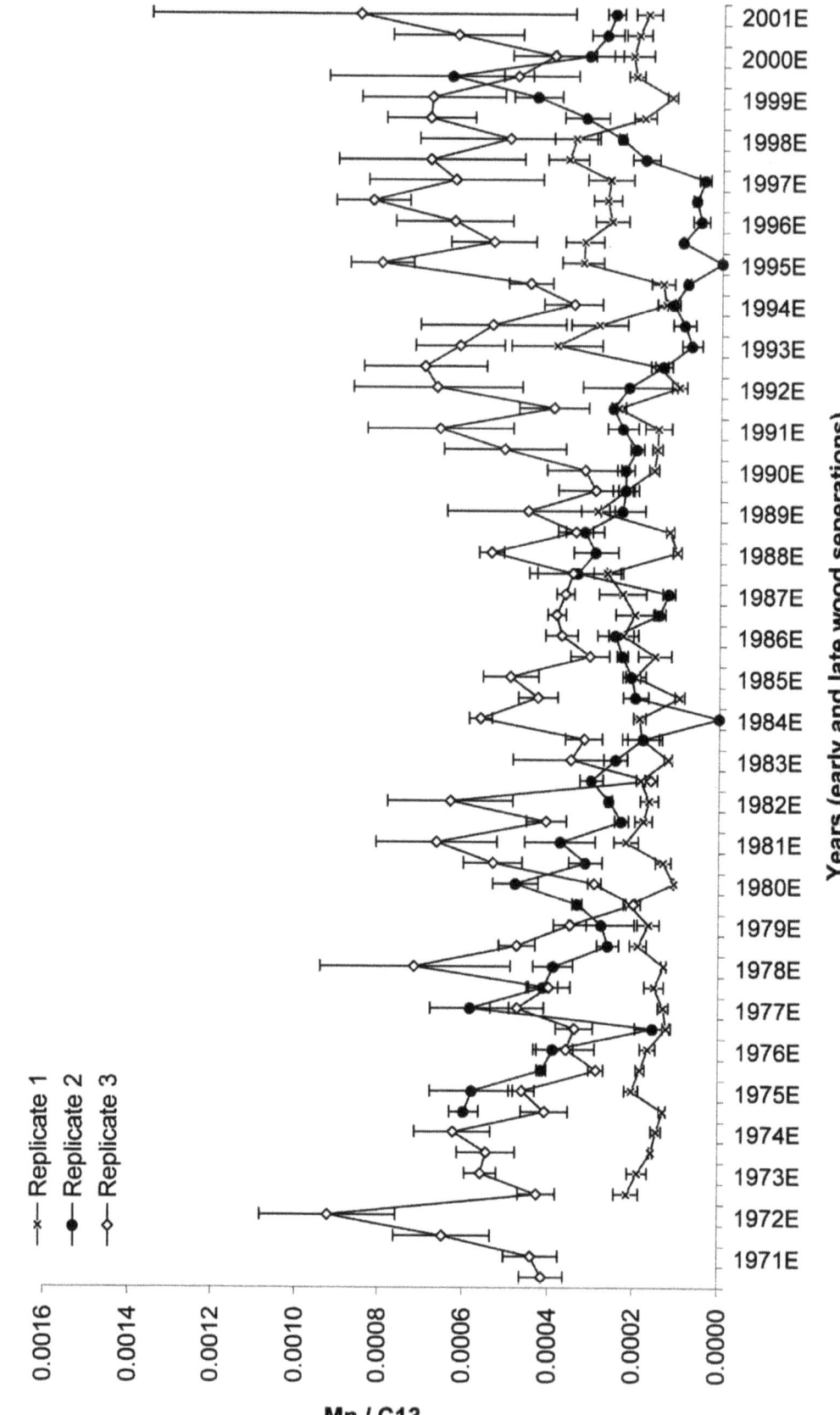

FIGURE 3.9: Three replicate runs of the same sequence of *Pinus contorta* for Mn

TABLE 3.5: Percentage standard deviation of ten ablations from a single tree ring for two wood samples and one homogeneous NIST disc (selected elements only). Where 'nd', the element was not determined for that sample

Element	Percentage Standard Deviation		
	NIST 610	Quercus robur 1	Pinus sylvestris 1
Al	13	21	23
Ca	13	25	14
V	13	19	24
Mn	9	15	55
Ni	12	19	17
Zn	14	22	15
Cu	5	55	12
Rb	11	15	32
Sr	6	29	19
Nb	5	28	nd
Cd	13	14	23
Ba	2	23	16

highly variable. Also, that prior to drawing any conclusions on data sets generated by the ablation of annual sequences of tree rings, a thorough investigation should be made of the levels of variation inherent in a selection of those rings, to estimate the overall reproducibility of the data.

In addition to this, an attempt was made to calculate how many ablations would be required to reduce the observed uncertainty to an acceptable level. This was based on the idea that this could be calculated mathematically by extrapolating how many samples would be required to bring the standard deviation down to a similar level to that found for NIST for a particular element. Using the following equation:

$$\sigma = \sqrt{\frac{\Sigma(x - \bar{x})^2}{(n-1)}}$$

where:
σ = sample standard deviation
x = any individual value
$\bar{x}$ = mean value
n = number of samples

A curve was produced from which it was anticipated it would be possible to read the number of ablations required to lower the error to a satisfactory level. By multiplying the number of ablations required by the volume ablated, it would have been possible to tell not only how big an area of tree ring would be required for a representative sample to be taken, but also, the volume of wood required for analysis. This would have had interesting implications for sampling design procedures and the overall use of LA-ICP-MS for the analysis of wood. However, for this to have been possible, data from the laser ablation of a single tree ring would have to display a normal distribution. As the wood is so heterogeneous, the data sets produced display a skewed distribution pattern (see figure 3.10 for an example). As a result, the curve generated by the equation plateaus out to infinity and so can not be used in the desired way. The implication of this is that unless the wood type being studied is especially homogenous, or some adjustment can be made to the way in which the laser ablates the wood (for example by defocussing to sample a far larger quantity of the tree ring) it will not be possible to use LA-ICP-MS for the successful analysis of inter-annual patterns of elements in tree rings.

As if all this was not negative enough, continued trial analyses of various wood types by LA-ICP-MS showed that of around fifty potentially detectable elements, twenty or less were regularly detected. Those that were routinely detected were Al, Ca, Titanium (Ti), V, Mn, Fe, Ni, Cu, Zn, As, rubidium (Rb), Sr, Y, zirconium (Zr), Ag, Cd and Ba, with La, Nd, or Sm occasionally above detection. The problem for this particular project is that the potential to finger print any volcanic trace is likely to lie with the rare earth, or other unusual elements. This fact does not rule out the potential to in-

TABLE 3.6: Percentage standard deviation of ten ablations from another ring from each of the same samples as table 3.5. Where 'nd' values could not be reliably determined for a particular element.

Element	Percentage Standard Deviation	
	Quercus robur 2	*Pinus sylvestris* 2
Al	14	10
Ca	9	12
V	nd	29
Mn	17	53
Ni	nd	9
Zn	12	8
Cu	42	17
Rb	21	28
Sr	nd	10
Nb	9	18
Cd	15	99
Ba	nd	18

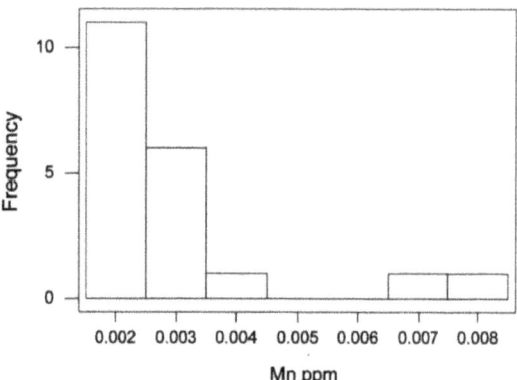

FIGURE 3.10: An example of the type of skewed distribution displayed by a typical element from a typical set of analyses from a single tree ring.

vestigate the possibility of proxy volcanic signatures (possible responses to increased environmental acidity in relation to relative patterns of certain essential elements) via LA-ICP-MS. However, this, coupled with all the previously discussed problems, clearly demonstrated that LA-ICP-MS, despite very obvious potential for the analysis of wood, at present does not produce data which can be much used to further the research question behind the project.

It was decided that the best option was to abandon LA-ICP-MS for the better established solution induction ICP-MS - a technique successfully used in other dendrochemical studies for the multi-elemental analysis of wood down to the rare earth elements [58]. It was hoped that the homogeneity of the tree rings might be countered by amalgamating several grams of material from an individual tree ring for each sample. Previous analysis had shown that the technique was capable of detecting a far greater range of elements in different wood types and with far lower error than the laser ablation technique. Whilst other research groups had analysed tree rings by ICP-MS, few had attempted it at annual resolution and with a primary aim of investigating rare earth element patterns. The main area for methodological development in this case was in the area of sample preparation.

3.4 Methodological Design: Solutions ICP-MS and ICP-AES

The methodological design for the cleaning, dissection and dissolution of individual tree ring samples for solutions analysis was a gradually evolving process throughout the project. Ideas taken from the literature were shaped by the various constraints of the project in terms of finance, available facilities and time. More crucially, new practical applications were developed and fused together with variations on old ideas to produce an overall new methodology for producing wood solutions for ICP-MS or ICP-AES analysis.

3.4.1 Sample preparation: Dissection

The development of an effective procedure for the accurate dissection of individual tree ring samples, minimising contamination, to produce a homogenous average sample of a particular year, has been based on trail and error. Every wood type dissects in a slightly different way, therefore step one of the developed dissection procedure was to examine the structure and hardness of any given sample under times twenty magnification. Based on this, decisions could then be made as to the optimum thickness of sub-samples and the best angles from which to dissect the tree rings. It is also important at this stage to be clear on the increment boundaries delimiting a single ring, and to be aware of possible structural inconsistencies such as fake rings and how to identify them in the context of the wood in question. Where necessary anatomical slides were produced and reference collections consulted. Large samples were then reduced into smaller sub-samples for analysis. The sub-samples were cut in cross sections at right angles to the ring pattern. The thickness of a typical slice ranged from 0.6 cm for soft, cleavable wood like *Pinus* sp., to 0.3 cm for harder woods such as *Quercus* sp. As a basic rule, the harder the wood and the more complex the anatomy the thinner the slice required for accurate dissection down the ring boundary. These sub-samples were then cleaned.

A variety of techniques were tried for cleaning the samples, these included sanding (with vacuum removal of the excess sawdust), the removal of the upper sample surface by shaving with the dissection blade, and washing with 20% nitric acid. Sanding was observed to introduce two forms of contamination, firstly from the paper itself and second in that the process spreads cells from various years and compacts them in the upper surface of other years. This was demonstrated by colouring one tree ring in a cross section with red dye, then sanding tangentially across the sample surface, it was observed that the red material was spread over the surface of the sample. For this reason the majority of samples were cleaned by removal of the the surface layer using a sterile steel blade of the same type used for dissection. Powder free, disposable latex gloves were worn throughout the cleaning process, these were exchanged for new ones during dissection. Whilst very time consuming this procedure was successful in that it provided clearly visible tree rings for dissection and replaced numerous possible contamination sources with one steel blade of known chemistry. Acid washing was used on occasion for harder woods, however the introduction of moisture to softwoods in this way was potentially very destructive as it was found to lead to the growth of mould. Any samples contaminated in this way were discarded, as various types of wood mold were tested and found to have higher concentrations of various elements than the wood itself. Acid washing was also found to fail to remove minute pieces of chain saw metal frequently found embedded in sample surfaces.

The best method for accurate and effective dissection was found to be application of pressure to a steel blade to split (or slice where necessary) down the natural planes of weakness along ring boundaries. For species such as *Pinus* with strongly defined boundaries between years and between the early and the late wood this technique was used to best effect with minimum contact required from the blade, thus reducing contamination. All dissection was carried out under magnification as necessary to ensure the best precision. Figure 3.11 shows illustrations of the various cleaning processes and the limitations of dissection.

The maximum available wood was sampled from each tree ring from as many different heights and angles of circumference as possi-

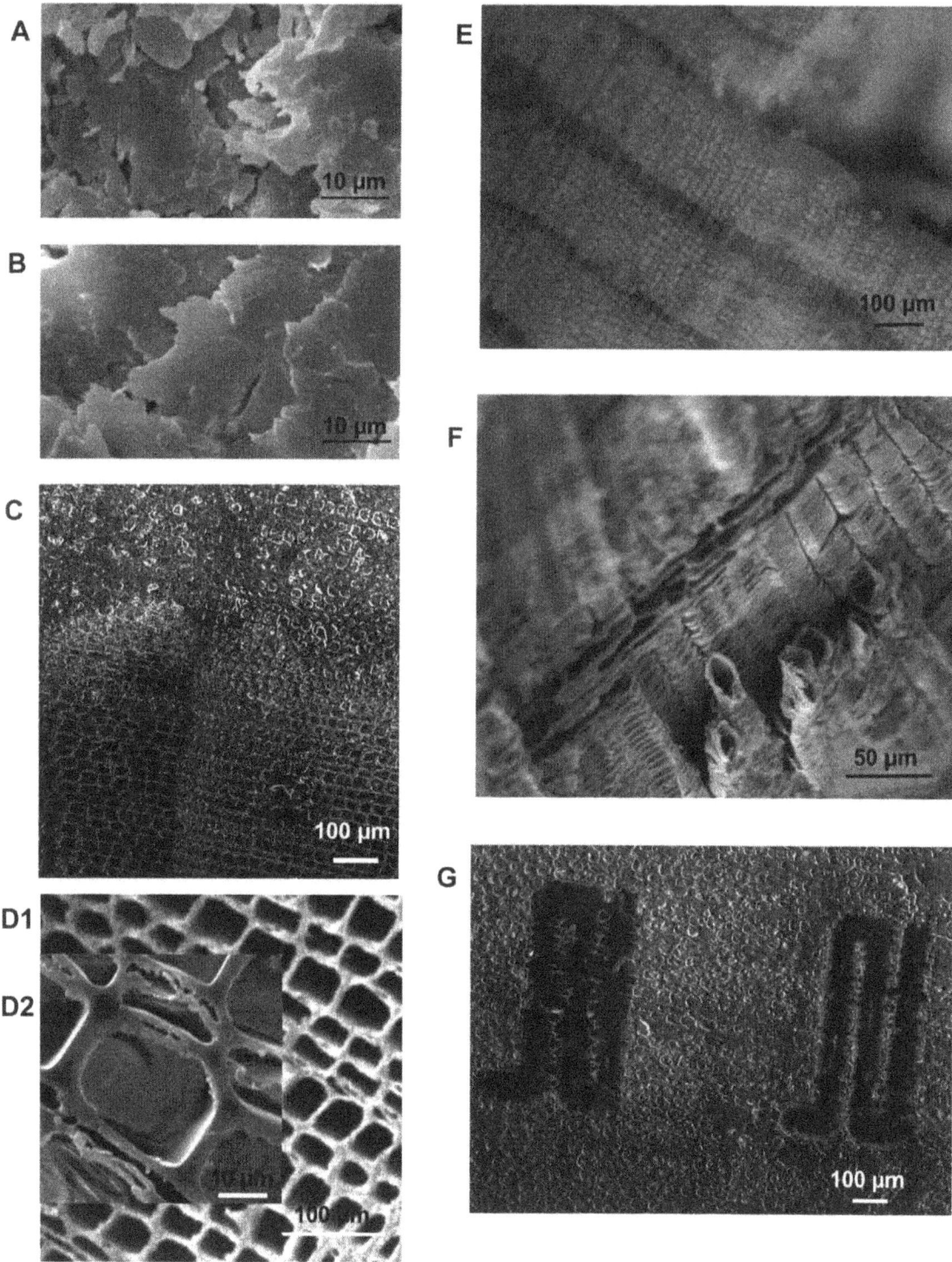

FIGURE 3.11: A: Sanded sample surface (note loose material) B: Sanded, vacuumed, acid rinsed sample surface (no loose material but cells smudged from one ring to another) C: Overview of a test sample (cleaned, sanded top, bottom surface cut with a steel blade) D1 and D2 freshly cut surface exposes just the cell structure for analysis E: Cleaned *Pinus* sample during dissection (to illustrate the type of precision achieved) F: Cleavage down the late / early wood boundary G: Marks left after laser ablation sampling of sanded wood

ble. The bulk dissected material for each sample was then chopped into a finer fraction with the dissection blade or ground down to more homogeneous fibres with an acid washed pestle and mortar. Once thoroughly mixed and homogenised in this way, a given amount of each sample was sub-sampled for dissolution.

Methods such as grinding the samples to sawdust (as used by some other dendrochemical studies) were rejected as it seemed likely that metal grinding parts might introduce too much contamination for the low blank requirements of this study.

3.4.2 Sample preparation: Dissolution

The ideal methodology for bringing wood samples into solution would involve an easily reproducible process, which would preserve all elements of interest whilst bringing the total sample into solution. It would involve as few steps as possible and as few reagents as possible in order to reduce contamination potential, and would be cost effective, making the best use of the facilities available.

A main aim for this step of the procedure was to produce the lowest blanks possible so that the maximum number of elements could be detected, this was especially important in the context of aiming to detect the rare earth elements. In an ideal scenario all samples would have been prepared in specially purchased vessels in a clean (purpose built, low contamination lab), designed especially for trace element work. However given the financial constraints of the project this type of facility and equipment were not always available.

Nitric acid was selected for disintegration of the samples as it is a strong oxidising agent which is known to liberate trace elements as soluble nitrate salts from biological materials. The nitric acid spectrum is similar to plain water which produces a low interference matrix ideal for ICP-MS. It is also available in a high purity form at a reasonable cost and is easy to handle once dilute (a consideration as completed samples were transported from the University of Reading to the NERC facility at Kingston-Upon-Thames for analysis).

Various options for digestion were tried, these included open boiling, microwave digestion and dry ashing. All were found to have various degrees of success for particular wood types, and advantages and disadvantages in the context of this project.

Open boiling is a cheap and easy method of bringing wood into solution and was found to work well in terms of wood types which were low in resin. 0.2 g of wood were placed in 2mls of concentrated ultra pure nitric acid and boiled on a hot plate for a number of hours. However for many *Pinus*. sp samples a cloudy precipitate was observed to form after dilution making the samples unusable. It was also necessary to evaporate the samples to incipient dryness prior to dilution in order to remove the majority of the nitric acid, this process was difficult to regulate and resulted in a proportion of samples being lost per batch. Open boiling can lead to the loss of volitiles such as mercury (Hg), As, Cd and Pb, although is known to be suitable for the quantitative determination of Ba, beryllium (Be), cobalt (Co), niobium (Nb), Ni, Pb, Rb, antimony (Sb), tin (Sn), Sr, tantalum (Ta), Ti, thorium (Th), uranium (U), W, Y, Zn and the majority of rare earth elements.

Microwave digestion in closed vessels is a preferable means of sample dissolution as most volatiles are retained in the process and there is less opportunity for contamination of samples. In a microwave, polar molecules and ions are energised by dipole radiation and ionic conductance. This results in rapid heating of the aqueous phase and the increased kinetic energy of the matrix increases dissolution rates over those created by elevated temperature alone [71]. The performance of acids used are also enhanced at the higher pressure engendered by the use of a closed vessel. Nitric acid for example, boils at a much higher temperature significantly increasing the oxidisation potential. The main draw back for this project was in terms of cost, even purchasing a set of vessels for use in a digestion microwave was beyond the budget of the project. The process for using a microwave is also very labor intensive, with only 10 samples (plus two blanks) being digested at one time, and an intensive acid washing programme of cleaning following between batches. Given the advantages however, the possibilities of using a digestion microwave were explored at the NERC facility at Kingston upon Thames. A batch of around forty samples of *Quercus* sp. (Kingston code 7618041) was prepared via microwave digestion according to the standard sample preparation note for wood pulp, issued by CEM (see Appendix C for details). The main problem with this procedure is that it resulted in a digested solution that contained 0.5g of wood in 10mls of concentrated nitric acid. This had then to be diluted down to ensure it was safe to run through the instrument (less than 2%). However, such dilution can result in the loss of rare earth and other elements at low concentrations. In order to the reduce the acidity without over diluting the sample, experiments were tried to evaporate off the acid fraction and re-dilute with ultra pure water. As mentioned previously this process is hard to regulate and resulted in increased opportunities for the contamination and loss of samples. Whilst microwave digestion is good for the quantitative determination of many elements including Ba, Be, Co, cesium (Cs), Nb, Ni, Pb, Rb, Sb, Sn, Sr, Ta, Ti, Th, U, W, Y, Zn and most rare earth elements, the methodology was abandoned in favor of a simpler, more cost and time effective alternative. With more time and further funding it is suggested however that this methodology could be developed into the method of choice for the digestion of wood samples.

The final methodology adopted for bringing samples into solution was a combination of dry ashing and nitric acid dissolution. 0.2g of sample was weighed into purpose bought 20ml Pyrex beakers which were placed in a furnace. A plan was made of the positions of the beakers within the furnace and samples were then ashed at a pre-determined temperature for a pre-determined length of time (these variables varied from wood type to wood type). Samples were removed from the furnace and left to cool in a desiccator for thirty minutes. 1ml of concentrated nitric acid was then added and the samples were agitated. The sample solutions were made up to 20mls with ultra pure de-ionised water and transferred into purpose bought autosampler tubes, samples were then cold stored until analysis.

Between batches of samples, the beakers were acid washed in boiling, 50% nitric acid for 4 hours, they were then rinsed five times in ultra pure water before being left to dry in a drying cabinet. The benefits of this simple methodology were that it was easily to replicate, extremely cost effective, produced suitability low blanks and in the majority of cases resulted in a complete digestion of the sample. In cases where the wood in question was extremely resinous or where some aspect of the preservation had altered the fibres, this digestion method was not sufficiently effective and minute particles of undigested material could be observed. In these cases a filtration step was added to the procedure and samples were passed through Watman 45 ashless filter papers in acid washed glass filter funnels, into new autosampler tubes. Attempts were made to quantify and correct for the loss of material from the overall sample by weighing the filter paper before and after filtration (once dry), however in the majority of cases the lost material was literally a couple of fibres and the weight was not possible to measure. Other disadvantages of the method were that it was still relatively time consuming and slight inconsistencies in temperature within the furnace could sometimes result in incomplete digestion. These problems could be overcome by the use of a larger, more effective furnace.

Prior to the analysis of each new wood type, waste wood shavings generated by the cleaning procedure were tested to fine tune the temperature and length of time required for effective ashing and subsequent dissolution. However a basic reference collection of six different wood samples was initially set up via a series of ashing experiments. Two species of modern *Pinus* were selected along with two samples of *Quercus* and two samples of *Juniperus* one from a modern tree, one from an ancient tree preserved in a specific burial environment. Material was dissected as for sampling. Five replicates of each of the six samples were placed in a furnace at 500°C for 3 hours. (This temperature was selected as any higher temperature was likely to volatilise the elements of interest). The samples were then removed from the furnace and left to cool in a desiccator for thirty minutes. The samples were made up to 20ml solutions as previously described, and were carefully studied to see if any undigested material could be observed. This process was repeated adding another hour each time from four hours up to fourteen hours and for some samples (modern *Quercus* and old *Juniperus*), twenty to twenty-four hours until a reasonable degree of digestion had been produced. Table 3.7 illustrates the main results and was used as a starting point for the ashing of all other samples, although each new sample achieved the

TABLE 3.7: Length of ashing period at 500°C required for total dissolution of samples. Results based on experiments with approximately 250 samples carried out over several weeks.

Sample	Hours Taken
Pinus halipensis	6 hours
Pinus nigra	5 hours
Quercus robur (Modern)	7 hours
Quercus robur (Old)	20 hours
Juniperus sp. (Modern)	14 hours
Juniperus sp. (Old)	24 hours

best results if the methodology was specifically tailored to that sample.

Differences in the ashing time for the two *Pinus* sp. were explained by the presence of a higher quantity of resin in the *Pinus halipensis* samples. It was found overall that less resinous samples ashed quicker and at lower temperatures to produce a total dissolution. The structure of *Juniperus* is similar to that of *Pinus* so it was logical that this sample was next to ash fully, with the more dense angiosperm structure of *Quercus* taking the longest of the modern woods. The results for the 'old' samples were more interesting as it is apparent that the environment of preservation played some part in the ease with which the samples could be broken down. In the case of the *Quercus*, the sample in question had been retrieved from a bog environment and was over a thousand years old, the *Juniperus* samples had been partly carbonised and buried in a building for over three thousand years. It seems likely that the structure of the *Quercus* sample may have been intrinsically weakened and / or enriched with organic material which resulted in an ash which could be completely dissolved after only seven hours. The *Juniperus* sample on the other hand was considerably harder to bring into solution, and even after twenty-four hours of ashing occasional samples were still found to contain minute fibres of undigested material which would need to be filtered out prior to analysis. This sample was especially problematic when it was required for ICP-AES analysis, as the temperature for ashing the samples was reduced to 400°C in order to prevent the main element of interest (sulphur) from being volatilised. At this reduced temperature it was not possible to achieve complete dissolution and even increasing the time the samples were left in the oven and adding a 'venting' step to introduce more oxygen to increase the effectiveness of the combustion process was not successful (for more details see chapter 6.1). Undigested material was filtered from the main samples and investigated under a scanning electron microscope (SEM), see figure 3.12.

Results showed that there was no preferential preservation of any part of the wood structure and that the undigested material was simply a collection of undigested fibres. It seems likely that the fact that the sample was already partially carbonised and, as is shown in chapter 6.1, contained high concentrations of some inorganic elements, were responsible for the difficulties experienced. Time did not permit the longer period of ashing which may eventually have turned the charcoal into ash, and so all samples were filtered prior to analysis. Given that ICP-AES has higher detection limits and can tolerate higher concentrations of elements, samples analysed could be more highly concentrated than those run by ICP-MS. Therefore, the difference made by this loss of material to the overall concentrations was calculated to be minimal in proportion to the total sample analysed. This also illustrates the impact of the nature of the environment of preservation on sample dissolution.

Only once procedural blanks for a particular data set had been run could a true idea of contamination introduced as part of the ashing and dissolution procedure be established. A standard blank of water and nitric acid was prepared immediately prior to analysis of a particular data set and analysed first. Next the procedural blanks (prepared at the same time and in exactly the same way as the samples) were analysed. These were followed by the calibration standards and one of the samples. The counts for the procedural blanks were compared with the standard blank. In an ideal scenario the standard blank and procedural blanks

FIGURE 3.12: A photograph of undigested, carbonised *Juniperus* fibres under a scanning electron microscope

would contain similarly low levels. It is usual for the procedural blanks to be slightly higher however, and the extent of this is the measure of how contaminated the samples are likely to be. In the worst scenario the procedural blanks were found to have counts of a similar order of magnitude to the samples for some elements. In this case sample batches would be re-prepared. In the majority of cases the procedural blanks displayed acceptably low levels of contamination, however one example of how the problem of blank contamination was managed will now be given.

For the first batch of solutions prepared for ICP-MS analysis it was noted that the raw counts for certain elements in the procedural blanks were high in comparison to the standard blank. It was clear that contamination was being introduced by some aspect of the sample preparation. Possible contamination sources were identified as the water and nitric acid used to dilute and digest the samples, the beakers in which the samples were ashed and digested and the tubes in which the samples were stored prior to analysis. With this in mind the water, acid, beakers and sample tubes used for the sample preparation procedure at The University of Reading (source A) were tested with the aid of duplicates from a source known to be low in contaminants - the NERC ICP-MS facility at Kingston-Upon-Thames, (source B). Blank samples were made up as follows:

- Tube B, water B, acid A
- Tube B, water A, acid B
- Tube A, water B, acid B,
- Tube A, water B
- Tube B, water B, acid B, rinsed round a beaker prior to addition.

These samples were duplicated. Results for selected elements are shown on table 3.8. By comparing element counts for the procedural blank, to the counts for the standard blank, and then to the counts from the A and B samples, the main sources of contamination were identified as the beakers and the auto sampler tubes. The biggest problem was Al which could be directly attributed to the tubes. More minor contamination with Zn and Sn was linked to the acid, and Sr and Rb to the beakers. Cu and Ca contamination came from several of the sources. A new bottle of acid and new low trace element tubes were used for all subsequent sample preparation. All beakers were boiled in 50% nitric acid, as described in section 3.4.2 as an effort to mitigate further contamination from this source.

3.5 Conclusions

An optimised field sampling strategy was designed for the collection of various types of sample in the field. This was based on experience over the course of the project and on a review of

TABLE 3.8: Results of the trouble shooting experiment to identify the sources of contamination introduced by the sampling procedure (source A is Reading University, source B is the NERC ICP-MS facility at Kingston-Upon-Thames). Data presented are raw counts of selected isotopes. As the data presented are raw data, the number of significant figures has not been rounded appropriately

Sample Blanks	^{27}Al	^{44}Ca	^{63}Cu	^{66}Zn	^{85}Rb	^{88}Sr	^{118}Sn
Mean of std blk:	13417	83069	199	539	27	91	41
Mean of procedural blank:	661614	100619	2066	8870	313	1685	157
A acid, B water, B tubes:	20057	85242	389	3597	44	187	176
A water, B tubes:	2633	73068	40	376	3	23	8
A beakers, B acid, B water, B tubes:	23138	85216	717	1319	118	606	48
B acid, B water, A tubes:	125738	76004	347	2229	58	142	63
B water, A tubes:	3053	70188	43	1174	34	56	11

the appropriate literature. It provides a starting point for further work, but in terms of this project was only put fully into practice on one occasion (see chapter 5) due to constraints on time and funding.

The main area for methodological design was in the use of LA-ICP-MS for the analysis of wood samples and the preparation of sequences of individual tree rings for analysis by conventional ICP-MS. The sample preparation methodology for LA-ICP-MS has not been previously considered in such detail and it is hoped that the simplistic measures developed as part of this chapter may be useful to others interested in the use of LA-ICP-MS on wood. The work carried out in terms of calibration clearly exposed the current weaknesses of the technique for the analysis of wood. Not only were attempts to produce precise and accurate concentrations data unsuccessful, but it was realised that underdeveloped calibration methodologies were compounded by heterogeneous tree ring chemistry. Further investigation into the degree of homogeneity possessed by tree rings, exposed an even greater problem in terms of attempting to gain well replicated data sets from sequences of heterogeneous rings. Without reliable replication of data sets generated from repeated analysis of the same sequence of tree rings, no findings from this mode of analysis could ever be substantiated. It was concluded that any potential future use for LA-ICP-MS in the context of dendrochemistry, would depend upon the tree rings for analysis being largely homogenous (e.g. *Acer saccharum* Marsh. [167]). The final difficulty exposed for the use of this technique on such complex biological samples, was the fact that it was not possible to detect a full range of elements due to the high background blanks. This meant that even if calibration problems could be over come, and a species of tree with relatively homogenous tree ring chemistry found for analysis, one of the main avenues of research for the project, namely to detect rare earth finger prints for volcanic eruptions, could not be addressed. It was decided in view of all these considerations, that the method of choice for further pursuit of the research question would by solution induction ICP-MS.

A cheap, relatively quick, but highly labour intensive methodology for the accurate cleaning, dissection and dissolution of samples of individual tree rings was developed and successfully put into practice (see chapter 5). A full range of elements were detected including many of the rare earth elements for the majority of samples. Precise and accurate concentrations in ppm and down to ppb were detected. The procedure developed principally for ICP-MS solutions was also used to prepare samples of archaeological material for analysis for S via the complimentary technique of ICP-AES (see chapter 6.1 for further details). ICP-AES can be used for dendrochemical analysis [6, 36, 46, 89, 153] and could be a useful complimentary technique where certain elements cannot be detected by ICP-MS. However in the case of samples analysed for this project, this was not proved, as it was concluded that the samples themselves did not reflect normal wood chemistry (see section 6.5 for further detail).

The following points summarise the outcomes of the methodological development in terms of the pros and cons of both ICP-MS methods:

LA-ICP-MS

- Pros

  1. Largely non-destructive, so can be used on rare samples which must be retained for future types of analysis.
  2. Requires only a small amount of sample material, so can be used on easily obtainable core samples.
  3. Can be used to sample at annual resolution in extremely narrow sequences of tree rings (from around 150 μm wide).
  4. Can produce semi-quantitative concentrations (relative to one another) for a restricted number of essential elements.
  5. Can be used for a rapid pilot analysis to get a general idea of sample chemistry prior to commencing a new sample batch.
  6. Can be used to explore variations in the chemistry of a single tree ring.

- Cons

  1. Can only be used reliably where tree rings are largely homogenous (e.g. *Acer saccharum* Marsh.).
  2. Difficulties replicating samples from the same tree or tree ring if the wood is particularly heterogeneous.
  3. Low level trace elements, and especially the rare earth elements are not readily detected.
  4. Precise, accurate quantitative data can not yet be obtained.

Solution ICP-MS

- Pros

  1. Produces data in absolute concentrations for a wide range of elements, including the rare earth elements.
  2. Can be used to sample at annual resolution provided tree rings are over 1000 μm in thickness, or where increments of several years are required.
  3. Can be used to take an average sample of largely heterogenous tree rings.

- Cons

  1. Samples are destroyed as part of the analytical process.
  2. Cannot be used for samples of annual resolution where tree rings are less than 1000 μm in thickness.
  3. A relatively large quantity of sample material is required for analysis.
  4. Sample preparation procedure is slow and highly labor intensive.
  5. Total digestion difficult to achieve in a cost effective manner which does not result in a loss of analytes.
  6. Numerous opportunities for sample contamination to occur.

The following three chapters report further results gained via one or more of the developed methodologies. Chapter 4 reports on a variety of small data sets gathered during the development of the various analytical methodologies, and their implications for the wider project. Chapter 5 puts into practice the optimised methodology for sampling, sample preparation and analysis of modern wood samples. Chapter 6.1 considers the complications of putting the same methodologies into practice in the context of the type of archaeological sample one might expect to encounter should the main research question be answered.

Chapter 4

Pilot Studies

4.1 Introduction

Over the course of this project, more than 2300 laser ablation samples, and 500 solution ICP-MS and ICP-AES samples were run. Of these, just over half produced what could truly be regarded as workable data. This was due to a combination of temperamental analytical equipment, a lack of specialist lab facilities for sample preparation, and the overall developmental nature of the research, with the tight quality controls which emerged. Aside from the two main case studies to be described in chapters 5 and 6.1, this section presents a selection of six of the additional pilot studies which helped shape the development of the project. These include data from a range of wood types, covering a number of volcanic events. Each study was initiated with a different set of aims and at various stages of the analytical development. They are included here partly to illustrate the attempted range and scope of the project and partly to report specific findings of interest to the project in general. Each study will be introduced in terms of the sample, the source of the sample and the main aims when commencing analysis. Important and / or unique) aspects of sampling and analysis will be discussed and a summary of the results will be illustrated with key graphs. Finally an outcomes section will describe the contribution of the particular study to the development of the project.

4.2 Pilot study 1

4.2.1 Introduction

Requests were sent to various overseas institutions for wood samples covering the period 1750 to 1850, and thus 1815, the year of the largest volcanic eruption of recent times, Tambora. If this event, recorded to have had a significant impact on global climate [126], could not be detected in the tree ring chemistry there would be little chance of detecting other more minor eruptions in wood ranging far from the eruption source. Samples of *Pinus sylvestris*, from Sarikamis, Turkey, were obtained from the Aegean Dendrochronology Project based at The Malcolm and Carolyn Wiener Laboratory for Aegean and Near Eastern Dendrochronology, Cornell, New York. The main aim for analysis was to identify whether a chemical trace which might be linked to Tambora could be found in the tree rings covering the event. Other aims were to explore differences in the chemistry between the early and late wood of each year, and to check the quality of the laser ablation data by preparing a sequence of tree rings for solutions analysis.

4.2.2 Sampling and analysis

A short test sequence from 1805 to 1818 (early and late wood) was run via LA-ICPMS. This was followed up by a 112 sample sequence of early and late wood from 1779 to 1835, and a further replicate sequence of 82 years of early wood ablations from 1779 to 1861. Samples were mounted to fit the dimensions of the laser sampling chamber and fresh sampling surfaces were cut prior to analysis. Laser sampling marks on the *Pinus sylvestris* sample can be seen in figure 4.1a. Data were calibrated to ^{13}C, and are presented as ratios to C. A sequence of 15 solutions, 1806 to 1820 were prepared. 3 mm thin sub-samples were separated from the main sample and cleaned under times twenty magnification with a pipette vacuum. Individual tree rings were dissected under times thirty magnification using a stainless steel razor blade. 0.25 g of each annual ring was placed in an acid washed boiling tube, with 5ml 69% concentrated nitric acid. The samples were left to stand for 24 hours, then placed in a digestion block and brought to a temperature of 140°C over a period of 3 hours, before being left to digest for a further 5 hours. Resulting solutions were transferred in to 10ml sterile auto-sampler tubes using three rinses of ultra-pure deionized water and made up to 10ml.

4.2.3 Results summary

The test sequence from 1805 to 1818 produced promising results. Figures 4.2 and 4.3 provide three examples. The early and late wood sequence for Al and Zn shows a marked heterogeneity around 1815. This is shown by the size of the error bars which reflect differences in the three runs of the single ablation which make up the sample. The larger the error, the larger the variation in tree ring chemistry only microns apart. The fact that the bars are small and stable for the rest of the sequence suggests that the chemistry of the tree rings is more homogenous for those years and supports the hypothesis that this becomes disturbed around 1815. The Cu graph shows a similar pattern, however the degree of error is such, that not only is inhomogeneity noted, but there is one significantly higher value in 1816, indicating an increase in the presence of this element. This appears even more convincing if just the early wood values for Zn and Cu are plotted (see figure 4.3). Here a true peak in the ratio to C can be seen for both elements. These results fit with hypotheses on the impact of environmental acidification (see section 2.7). The perceived rise in concentrations might relate to increased uptake of these elements in relation to an increase in environmental acidity. The heterogeneity may relate to the formation of the *hotspots* (see section 2.5.4) which are thought to be due to sudden, large scale mobilisation of divalent cations following acidic deposition on soils [163].

However, when additional extended replicate sequences were run from the same sample, with a view to testing the quality of the initial data, it did not prove possible in the given time to support the promising primary data sets. Figure 4.4 shows two samples of the original test sequence along with the two replicates for that part of the sequence. Al shows an example of one of the better correlations between replicates, Zn was one of the worst correlated replicates. Al showed a far closer replication between the second two analyses, than the first. There were several months delay between the first analysis and the second two, which were run one day after another. The fact that for most elements the degree of replication was better for the last two replication series, indicates that instrumental function and stability may have a large role to play in the failure to replicate sample sequences. However, the failure to detect the heterogeneity recorded by the first run is of more concern. Although this could be explained by suggesting that the second two runs failed to hit a *hotspot* within the significant tree rings, the fact remains that for any data set to be truly usable it must be possible to replicate the results. It should be considered that the heterogeneity detected in the first run might simply be coincidental, and caused by fluctuations in instrumental detection at the particular time those years were analysed. In view of the stability of the other error bars and the fact that the larger error bars do not occur for all elements at this time, it seems likely, however, that the fluctuations in chemistry are real. A further explanation is that, as has been noted in section 2.5.4, specific parts of tree ring anatomy can have a unique chemistry of their own. For example, investigation of a resin duct from the same sample showed that the cell structure of the duct contained more Ca, Mn, Zn, Cu, Rb, Sr, Ba and La than the normal cell walls. This being the case it was possible that the different runs of ablations hit different areas of the tree ring anatomy, however, microscopic examination of the ablation marks indicated that just normal cell structure was sampled for each of the hetrogenous years.

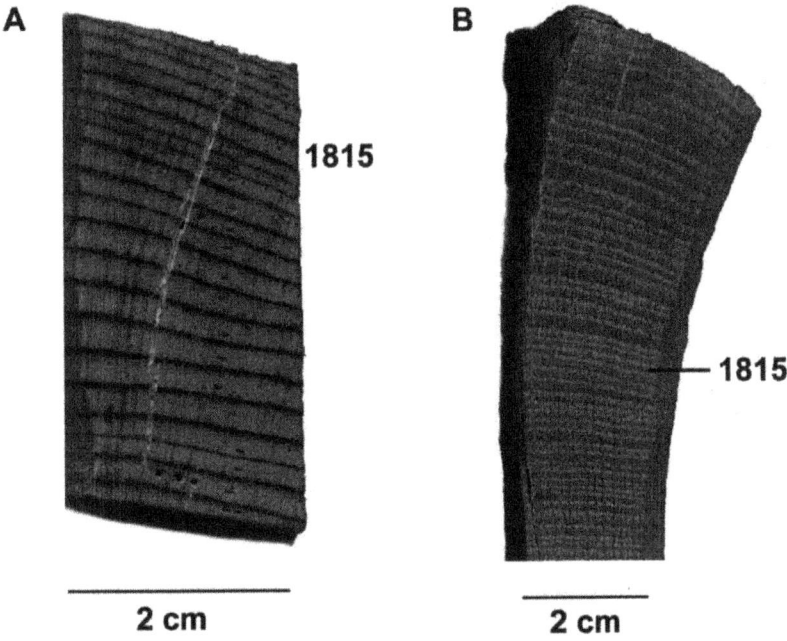

FIGURE 4.1: A: *Pinus sylvestris* sample from the Aegean Dendrochronology. Marks from the laser ablation process can be seen up the right side of the sample. B: *Quercus robur* sample from the Irish Chronology.

Although no convincing replication could be observed between the 1805 to 1818 repeated sequences, when the two extended sequences were plotted in full, several interesting patterns were observed. The best example is given in figure 4.5. Here Al and Ca data are shown for the first replicate sequence of early and late wood. Rises in these elements can be seen around 1783, 1800 and 1815. This is very significant as these dates coincide with major volcanic eruptions (Laki, Mt. St. Helens and Tambora). Other elements, e.g. Rb, showed spikes at these times, (1783, 1801 and 1815), however the sequence produced by the second replicate run was totally different. The early wood of the first run and the extended early wood run of the second sequence are shown for Al in figure 4.6. This final example shows that for Al at least this signal can be viewed as promising. Whilst this shows that there was some instrumental drift on the first day's analysis, the data for the second day show rises in the same parts of the sequence. A similar result was observed for Ca but the other elements showed less agreement.

The results of the solution analysis of the samples were found to be void due to what transpired to be a lack of clean lab facilities for sample preparation (contaminated glassware). Many of the samples also contained a precipitate and required filtering before analysis, showing the wet digestion block procedure to be unsuitable for future solutions preparation.

4.2.4 Outcomes

This pilot study produced the first promising results to suggest that it may one day be possible to detect certain large volcanic eruptions in rings from trees from various locations around the globe. The problems experienced with attempting to replicate the data sets, and the issue of the overall heterogeneity of tree rings (the examination of the resin duct and the differences found between the chemistry of the early and late wood) were key in shaping the progress of the project. The observations made in this study relate directly to sections 2.5.4 and 3.3.4. In terms of methodological development, the digestion block methodology was rejected for all solutions analysis, new glass ware was purchased specifically for the project and alternate lab facilities were used for subsequent sample preparation. For all future analyses special care was taken to ensure that only areas typifying the normal cell structure of the tree ring were ablated.

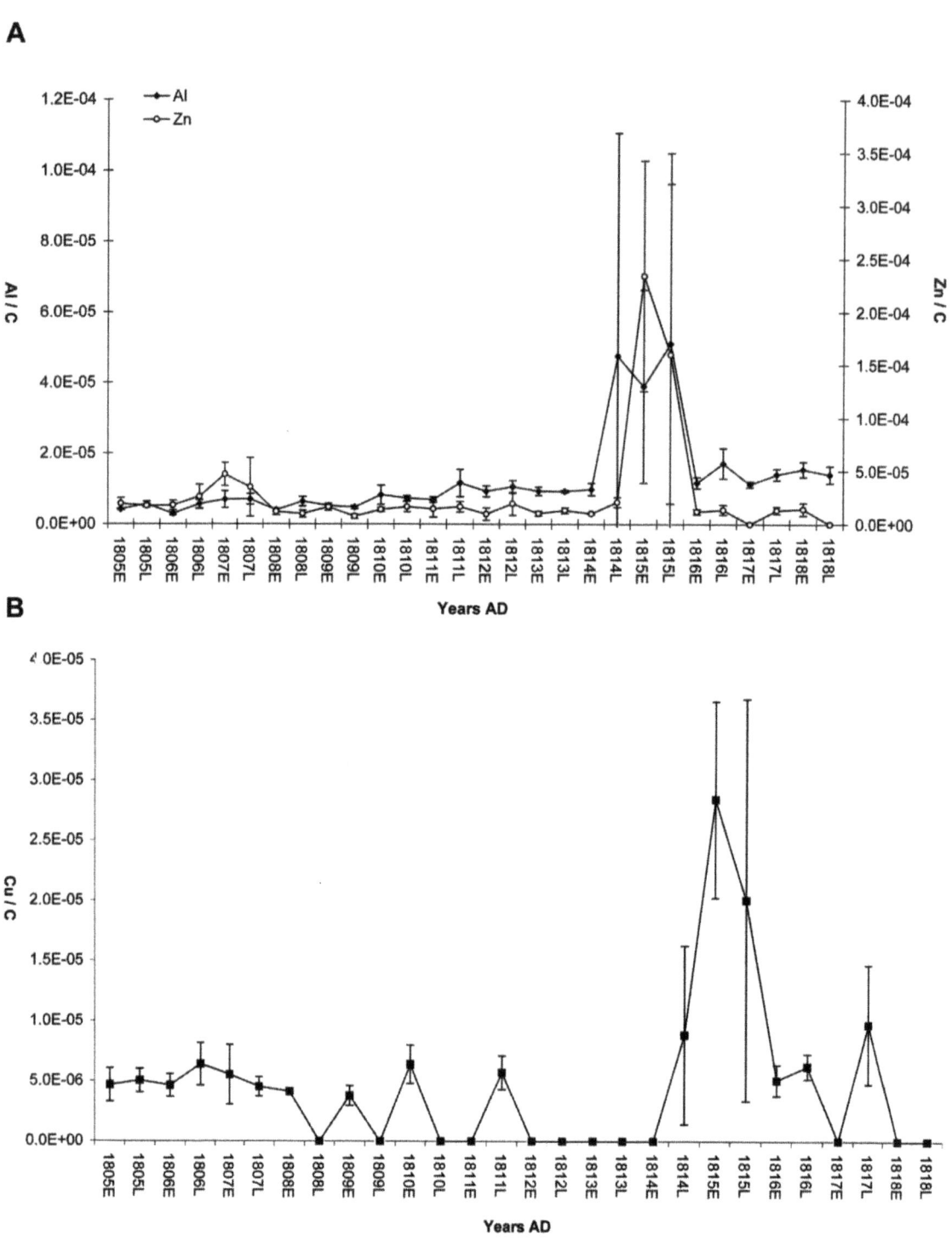

FIGURE 4.2: LA-ICP-MS analysis of *Pinus sylvestris* 1805 - 1818 AD. A: Early and late wood sequence for Al and Zn. B: Early and late wood sequence for Cu.

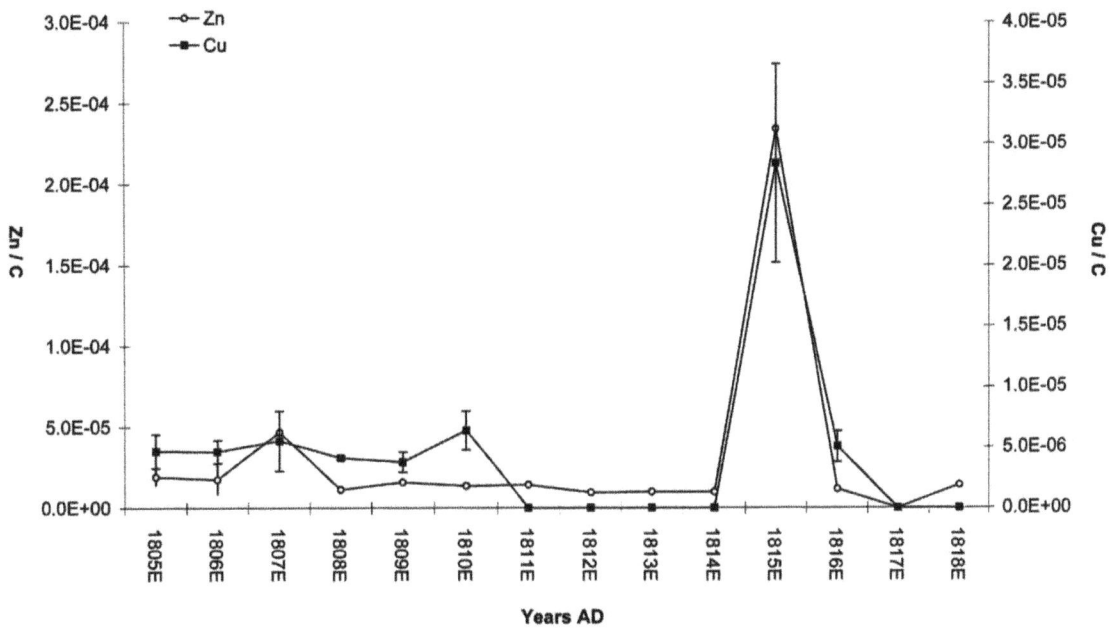

FIGURE 4.3: LA-ICP-MS analysis of *Pinus sylvestris* 1805 - 1818. Early wood sequence for Zn and Cu.

4.3 Pilot study 2

4.3.1 Introduction

A second sample covering the Tambora eruption, was obtained from from the Irish Dendrochronology, Queen's University, Belfast. The sample was *Quercus robur* from tree Q8513AA, Shanes Castle, Northern Ireland, and was terrestrially grown (as opposed to one of the bog oaks from which a large percentage of the chronology is built). The main aims for analysis were to identify whether a chemical trace which might be linked to Tambora could be found in the tree rings covering this event and to analyse a long sequence which investigated the chemistry of both early and late wood. It was also to be an investigation of the differences in conducting analysis on a complexly structured angiosperm species compared with the previously considered gymnosperm samples. It was hoped this research would contribute to the on-going debate considered in Chapter 2, section 2.6.2, as to whether angiosperms or gymnosperms are more suitable for dendrochemical studies.

4.3.2 Sampling and analysis

Two sequences were run via LA-ICPMS, using the preparation methodology outlined in pilot study 1. A short test sequence was run from the early wood of 1801 to the late wood of 1817. A second sequence of 118 samples was run, from the late wood of 1772 to the early wood of 1831. The sample is illustrated in figure 4.1 b.

4.3.3 Results summary

Of over thirty elements scanned for, only Ca, Mn, Cu, Zn, Rb, Sr and Ba were reliably detected for both sequences. Whilst no striking change in chemistry appeared to occur around 1815 for these elements, slight changes could be argued for Mn, Cu, Rb, and Ba. The replicated section of the sequence showed a degree of agreement for Mn. Figure 4.7 shows the complete sequence of early and late wood samples for Mn, for both sequences. The degree of replication shown by the sequence, error allowing, is reasonable (although by no means statistically significant). Even so, it is interesting to note that for both sequences the pattern appears to

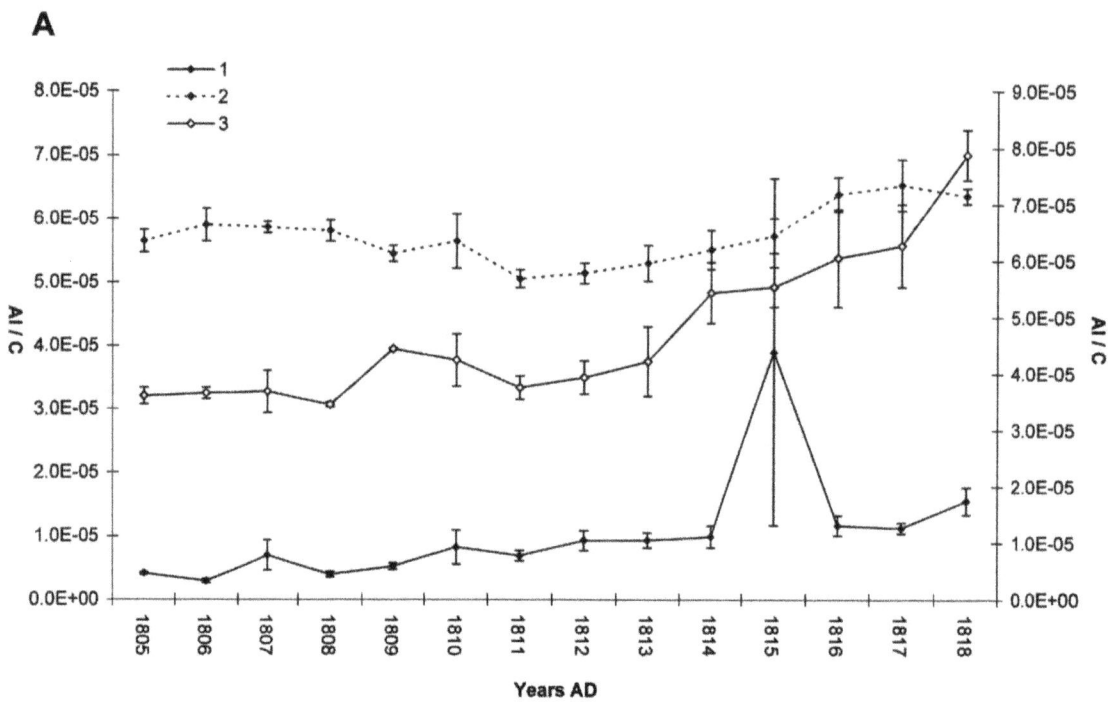

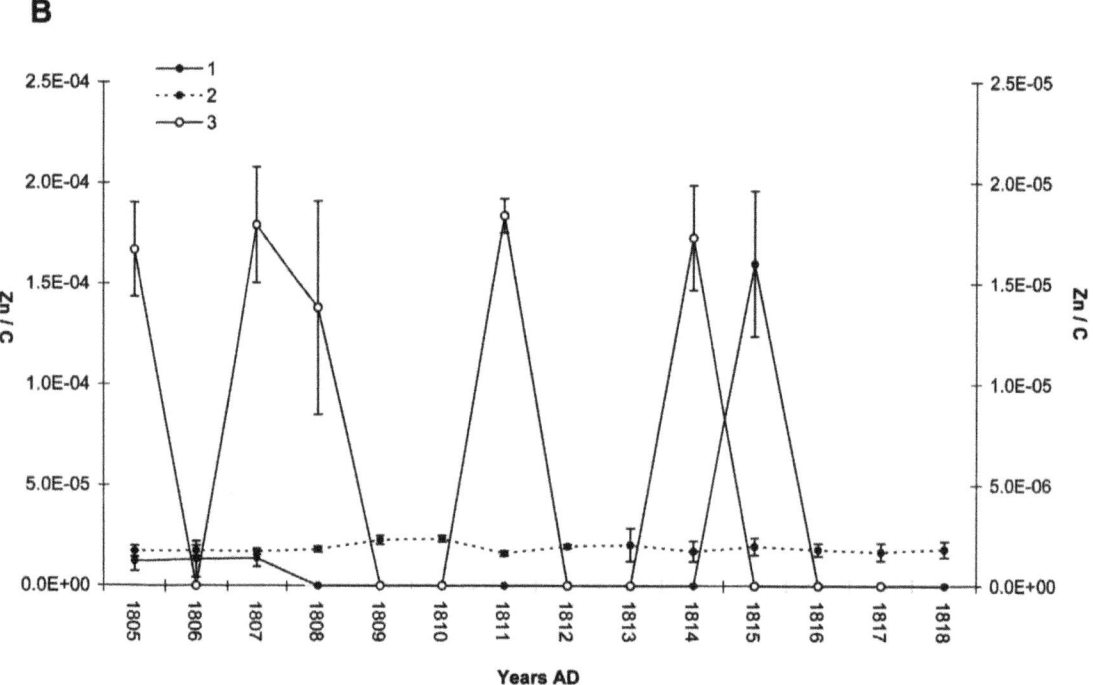

FIGURE 4.4: LA-ICP-MS analysis. A: Replicate analyses of Al in *Pinus sylvestris* early wood. B: Replicate analyses of Zn in *Pinus sylvestris* early wood.

Chapter 4: Pilot Studies

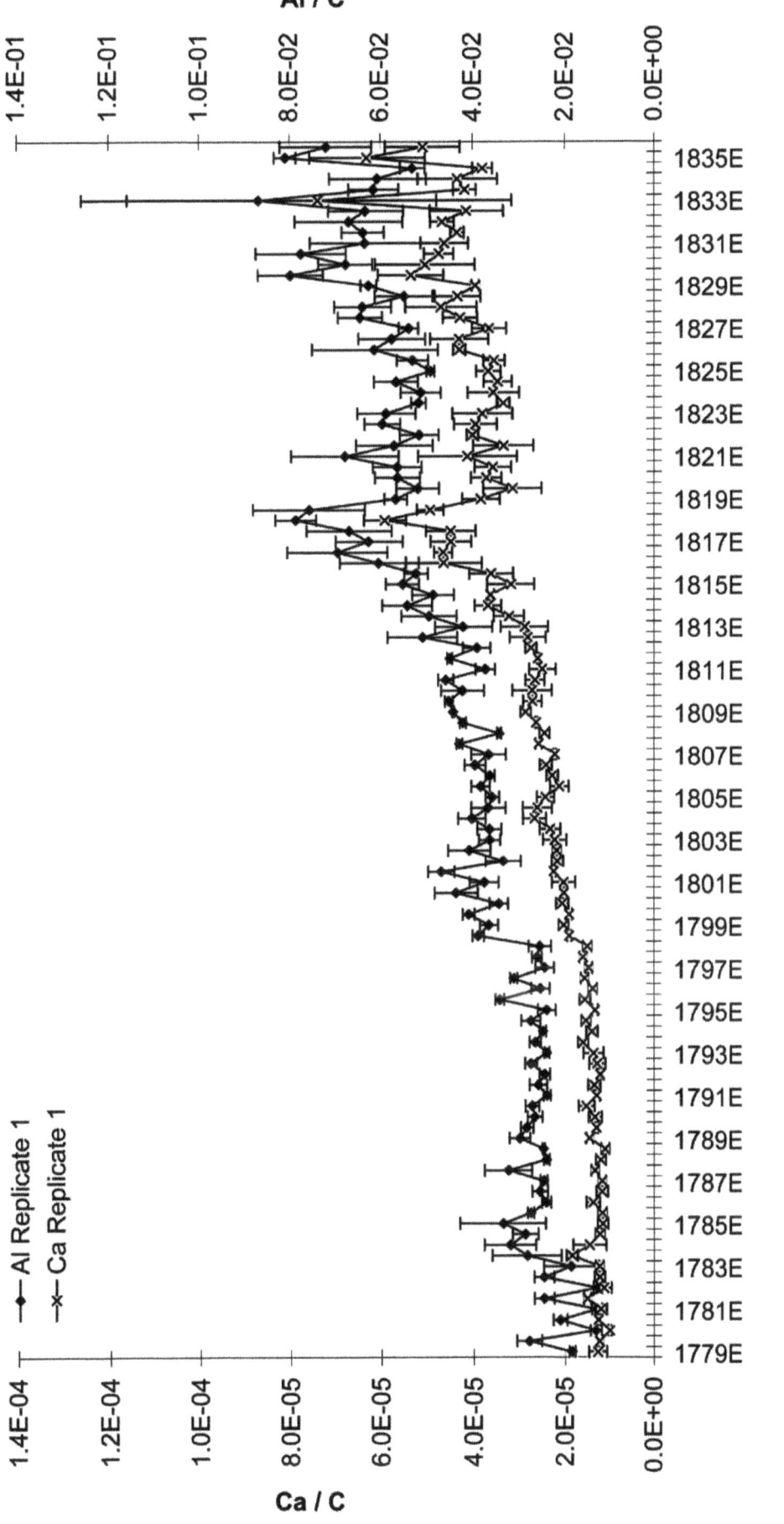

FIGURE 4.5: LA-ICP-MS analysis. Al and Ca data for replicate sequence 1, early and late wood (early wood only labelled on graph)

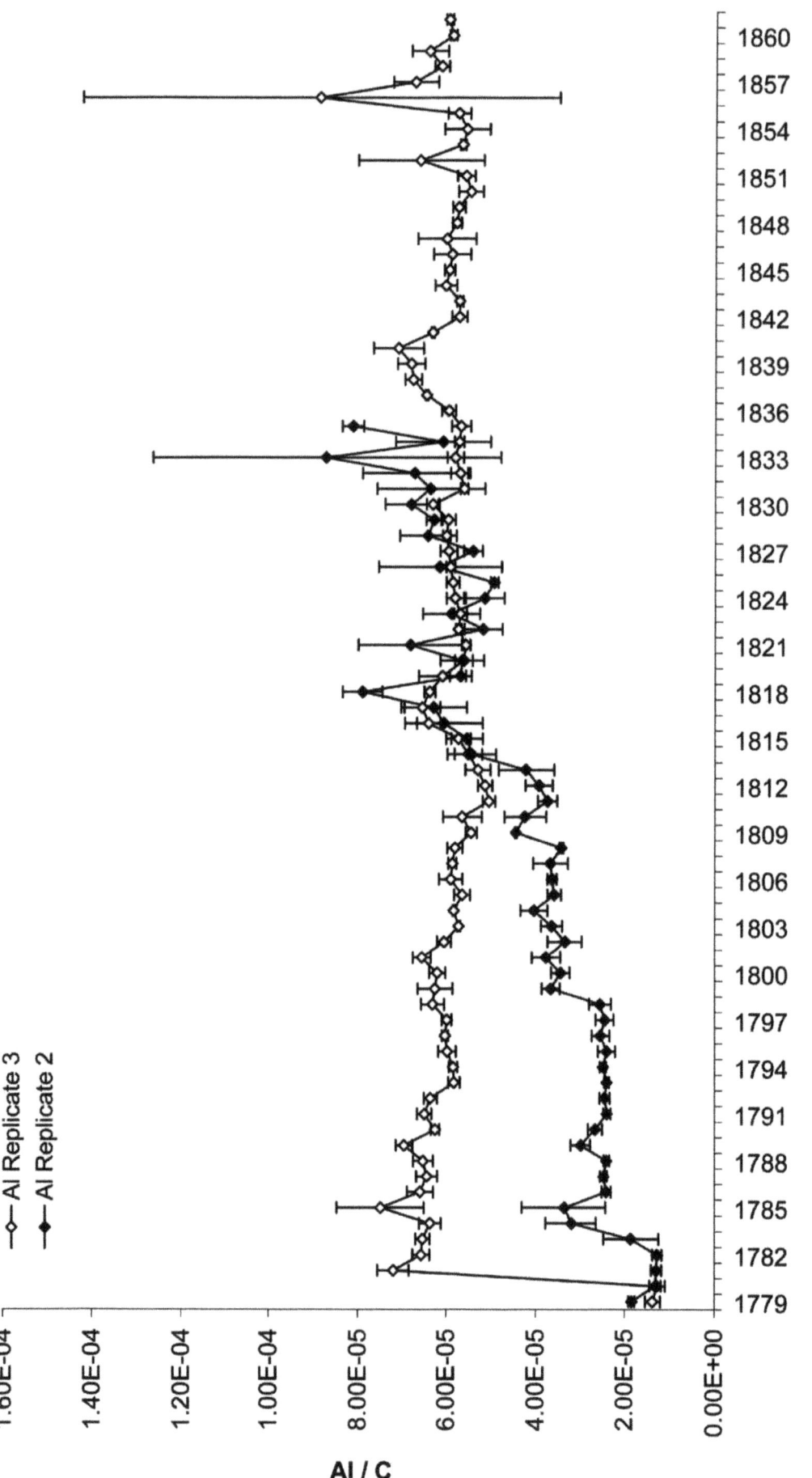

FIGURE 4.6: LA-ICP-MS analysis. Al data sequence replicated over two days of analysis. (early wood sequence only)

CHAPTER 4: PILOT STUDIES

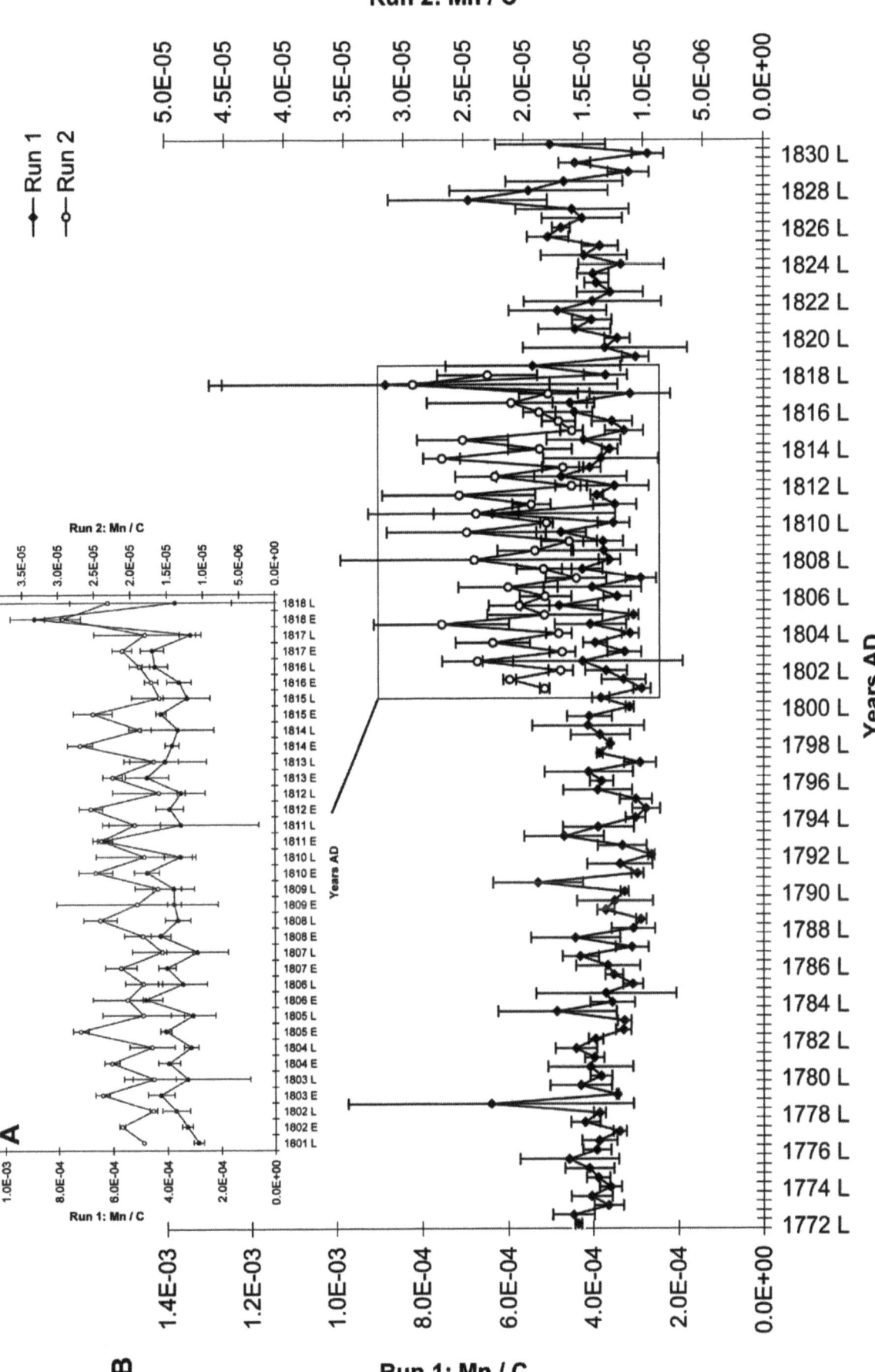

FIGURE 4.7: LA-ICP-MS analysis. Early and late wood sequence for Mn. A: Close up of the duplicated section of the sequence (1801 - 1817). B: Complete sequence (1772 - 1831).

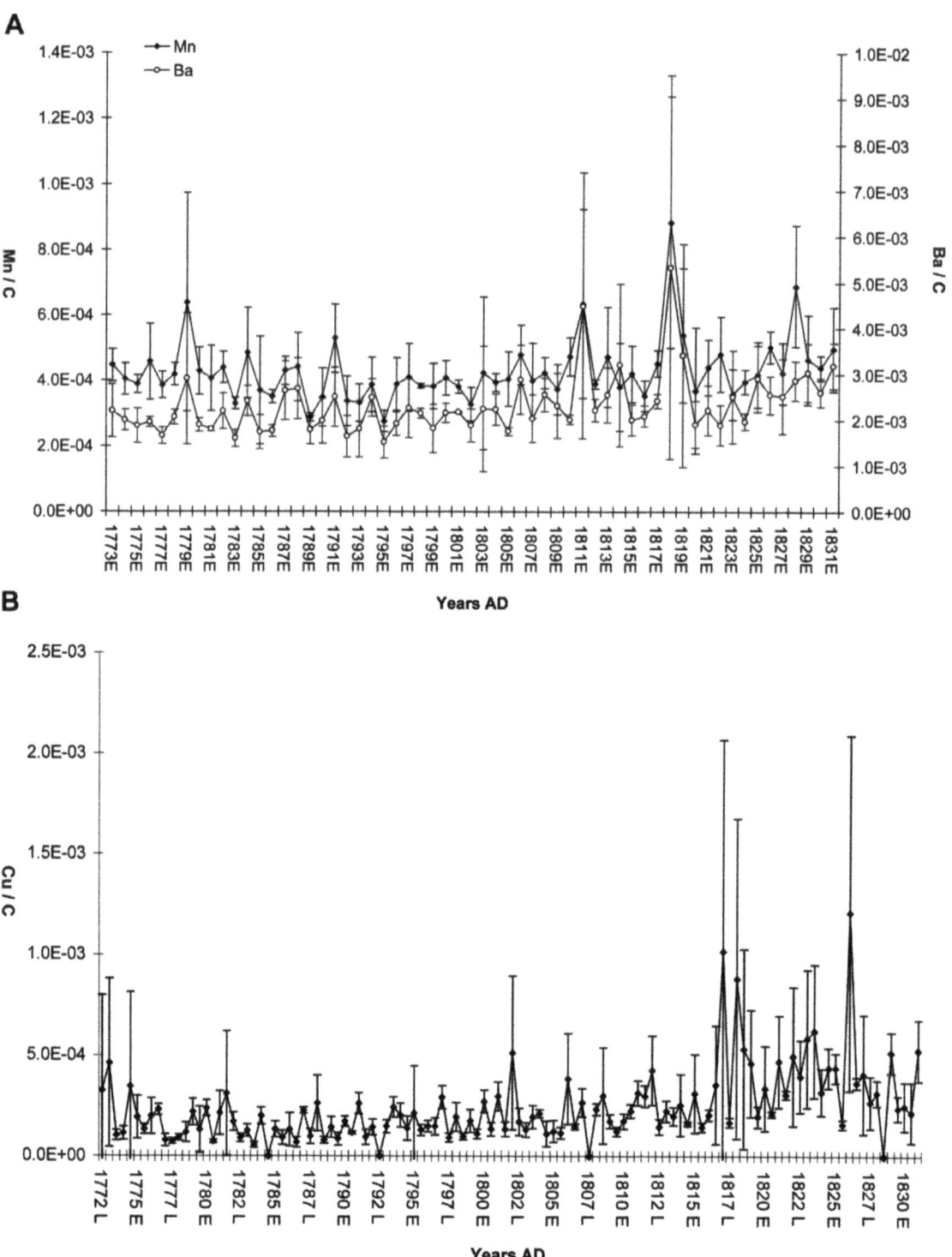

FIGURE 4.8: LA-ICP-MS analysis. A: Early wood ratios for Mn and Ba. B: Early and late wood ratios for Cu

alter between the 1815 late wood and the early wood of 1817. Both also include an anomalously high value in the early wood of 1818 (this also occurs for Rb). Placed in the context of the wider sequence, the 1818 value remains significant as it is still potentially higher than at any other point (subject to error), however the presence of a similarly high, and more defined peak in 1828, brings into question its significance in terms of a volcanic hypothesis. This figure also illustrates the dominantly lower ratios obtained for the late wood than for the early wood, a pattern observed for many elements in various *Quercus* sp. analysed. Figure 4.8 provides two further examples of increases in heterogeneity around the time of the Tambora eruption, and also at times of other major eruptions. Figure 4.8a, shows the agreement between Mn and Ba in the early wood. Three potentially higher ratios (and increased heterogeneity) can be observed for both elements in 1779, 1811, and 1818. Again, whilst tenuous, this could relate to increases in environmental acidity due to eruptions in these years (Sakura-Jima in 1779, an eruption of unknown origin around 1809 [26] and Tambora). Figure 4.8b, shows an increased homogeneity for Cu from 1816, which appears to be returning to more normal levels from around 1826 onwards.

4.3.4 Outcomes

This pilot study provided a useful comparison for study 1. The results showed some degree of agreement with the results for *Pinus* sp., however nothing truly conclusive in terms of tracing volcanic events. Exploration of the early and late wood chemistry produced an interesting insight into the variation within a single angiosperm tree ring. It was hypothesised that the lower ratios for the late wood could relate to lower functionality of the tree at the time of late wood formation, i.e. towards the end of the growth season. The higher values for the early wood would correspondingly relate to increased growth activity at the beginning and peak of the growth season. The complex angiosperm structure[1] of these samples made it difficult to take a representative sample of the cell structure via laser ablation. This may go some way towards explaining why, when tests were carried out to assess the homogeneity of the tree rings as a whole, this species was found to be particularly heterogeneous. In view of this, it was concluded that dendrochemical analysis, via LA-ICP-MS at least, was likely to be less successful on angiosperm species.

4.4 Pilot study 3

4.4.1 Introduction

Requests were made to various overseas institutions for wood samples which grew relatively near to certain volcanoes at the time of particular volcanic eruptions. Samples of *Agathis australis* (D.Don) Salisb. were obtained from the school of Geography and Environmental Sciences, The University of Auckland. These New Zealand samples included two cores from the Manaia Sanctuary, the Coromandel Peninsula, and a sub-sample of a large tree which grew close to Rotarua. The cores from the Manaia Sanctuary covered the Tarawera eruption of 1886 and grew within one hundred kilometers of the eruption site. The sub-sample of the larger tree covered the mid fourteenth century period when the Kaharoa eruption is thought to have taken place. The two main aims for analysis were to ascertain if any chemical changes could be detected in the tree rings in 1886 / 1887 associated with Tarawera, or to see if any signature could be detected somewhere in the mid-1300's which could be used to help pinpoint the date of the Kaharoa eruption. The overall objective was to see if a more distinct chemical trace could be detected in tree rings in close vicinity to particular eruptions.

4.4.2 Sampling and analysis

The samples are shown on figure 4.9. The core samples had been mounted and glued on to strips of MDF for measurement. Prior to sampling they were removed from the strips and the surface contacting the glue was removed with a steel blade. All sample surfaces were cleaned in the same way. Sub-samples covering the specific years of interest were taken from all the samples and re-mounted for the laser sampling chamber. Three sequences were analysed by LA-ICP-MS, with the aim of attempting to

---

[1] The typical cell structure of *Quercus* sp. can be observed in pilot study 4, figure 4.13, a and b.

find evidence of the 1886 Tarawera eruption. A pilot sequence from 1882-1890 was run from one of the cores, followed by a longer sequence from 1880-1900 from the other core. In addition, a fifty year sequence, 1330-1380 was analysed from the Rotarua sample in an attempt to locate the Kaharoa eruption, (1348 was missed during sampling due to narrowness of the ring pattern). Data were processed and presented in ratios to ^{13}C.

4.4.3 Results summary

Figure 4.10 gives examples of the most promising results gained for the test sequence (1882-1890) from core 1. Ca, Mn, Sr, Ba and Zn all showed their highest ratio in 1885. Ti values are only above detection in 1885 and 1886, and As is present in 1883 and declines from 1886 onwards. The occurrence of the highest quantity of certain elements in the year preceding a volcanic eruption can be explained via the process of translocation between the living rings of the sapwood. This conclusion was reached by Hall et al. [58] in reference to some of their rare earth element peaks in years preceding dated eruptions of Mt. St. Helens. The observed patterns could relate to increased environmental acidity. However, as can be see from figure 4.11, when an attempt was made to replicate an extended version of this sequence it was not possible to produce similar results from another tree from the same location. Therefore no chemical connection with the Tarawera eruption could be found.

The data for the fifty year sequence run with the aim of tracing the Kaharoa eruption were even less convincing. Of the seven or so elements reliably detected, the most promising pattern was produced for Mn, Sr and Rb. Each of these elements displayed a slight downturn after 1350 (see figure 4.12 for an example for Mn and Sr). Whilst this is very tentative, the observed decrease being only slightly more pronounced than those earlier in the sequence, the decrease and later recovery corresponds with an event which caused a narrowing of the ring widths around this time. If this pattern could be shown to be replicable, it might point to the fact that nutrient up take is less in years where growth conditions are more difficult for the tree. If this was the case, this evidence could be added to other proxy data to support 1350 as the date of the Kaharoa eruption.

4.4.4 Outcomes

This pilot study showed some promise, however, the failure to replicate any results from the Manaia Sanctuary trees, made the data and / or the suitability of *Agathis* sp. for dendrochemical work, questionable. One of the main problems with the study was that only six or seven elements out of the wide suite scanned were reliably above detection at any point in the sequences. This was concluded to be partially due to the chemistry of this species, but also due to fine tuning issues with the laser ablation technique.

4.5 Pilot study 4

4.5.1 Introduction

A pre-modern wood sample covering a controversial tree ring anomaly, possibly linked to the impact of a volcanic eruption of unknown date, was obtained from the Irish Dendrochronology. This *Quercus robur* sample was from tree Q9807, Deer Park, Northern Ireland, and had been preserved in a bog environment. It featured a narrow ring event, beginning 535/536 AD, from which the tree apparently never really recovered. This same event (typified by a clear negative signature around 536, followed by reduced growth and in some cases recovery some years later) was observed in fourteen out of fifteen separate *Quercus* samples from the Irish master chronology [8]. Figure 4.13 provides an illustration of the sample along with detail on the 535 / 536 AD event. This event has been linked by some to historic records from Europe and the Middle East of a mystery dust veil observed around this time [149]. Whilst Baillie [10] has suggested this dust veil may be attributable to near misses with comets, others have suggested it is more likely to be due to a volcanic eruption [77, 128]. The aims for analysis were to see if the chemistry of the tree rings around the event could be traced and / or to provide clues as to the origin of the event. However before this anal-

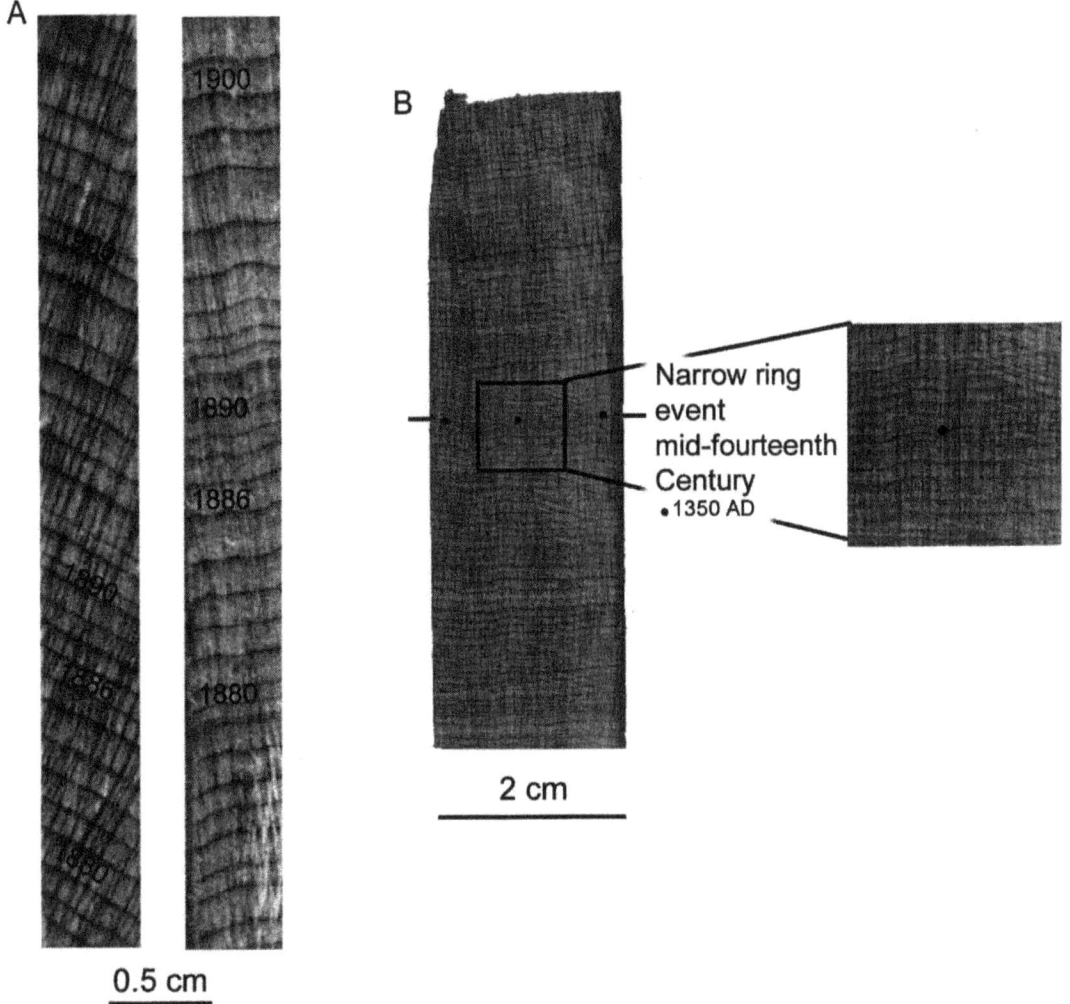

FIGURE 4.9: *Agathis australis* (D.Don) Salisb. samples. A: Cores from the Manaia Sanctuary covering the Tarawera eruption of 1886. B: Fourteenth century AD sample thought to cover the Kaharoa eruption.

ysis was carried out it was decided to investigate the general chemistry of the sample with the aim of determining the possible impact of preservation in the bog environment.

4.5.2 Sampling and analysis

In order to investigate the impact of the bog environment in which the sample had been preserved, ten years of the bog oak and ten years of a modern, terrestrially grown *Quercus* sample were dissected and brought into solution, the samples were then analysed by solution ICP-MS. In addition, a 50 year sample sequence of 25 years of early and late wood covering 518 to 543 AD was run via LA-ICP-MS.

4.5.3 Results summary

The average concentrations of the various elements in the two wood samples were compared. 52% of elements detected were higher in the bog oak than in the terrestrially grown oak. Where values were considerably higher, the presence of the element in question was concluded to be due to contamination in the bog environment. Table 4.1 outlines what appear to be the main contaminants. The rest of the values were very similar or lower in the bog oak compared with the terrestrially grown sample. It is possible that patterns created by these elements may represent reactions to environmental change at the time of tree ring growth. However, with

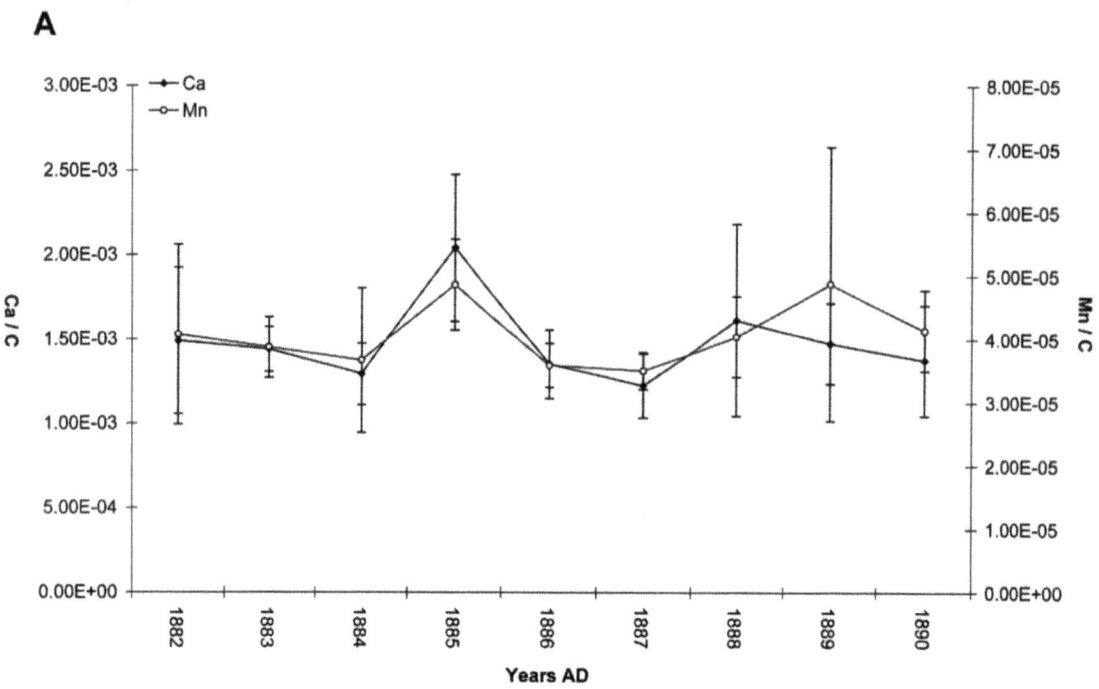

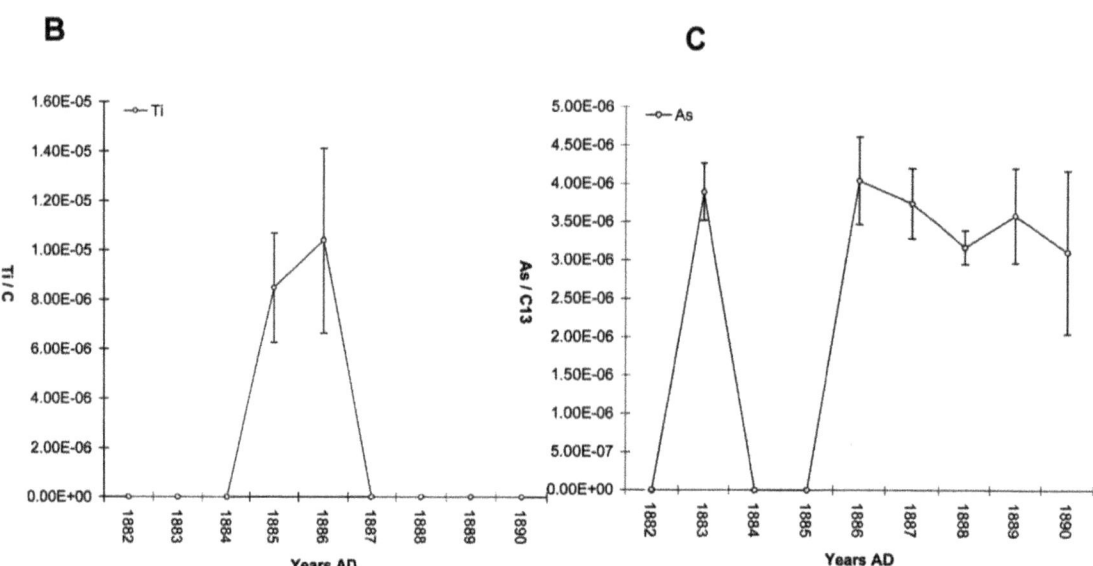

FIGURE 4.10: LA-ICP-MS analysis. Examples of results from *Agathis australis* test sequence 1882-1890 AD. A: Ca and Mn; B: Ti; C: As.

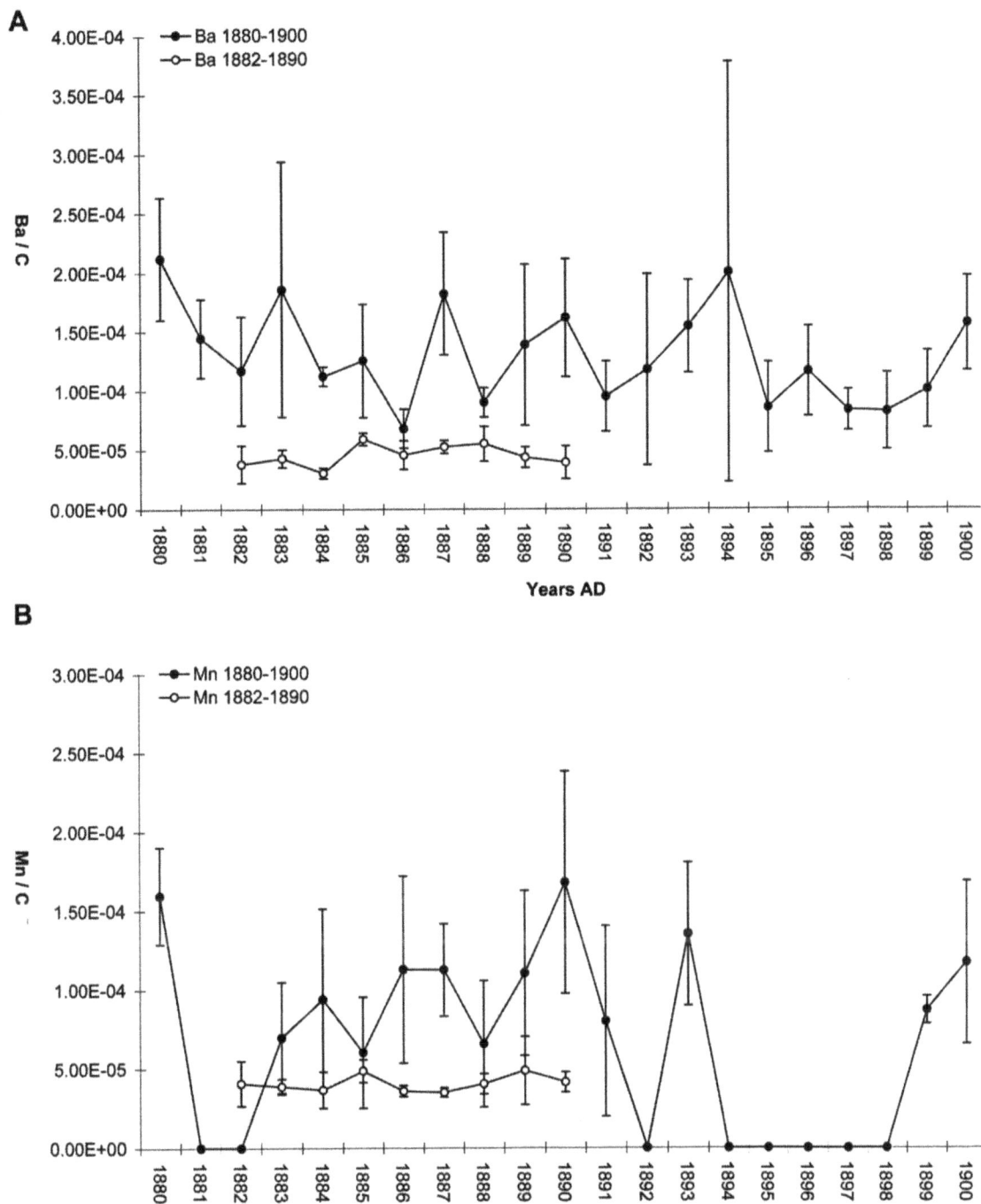

FIGURE 4.11: LA-ICP-MS analysis. Examples of results from *Agathis australis* test sequence 1882-1890 and 1880-1900 AD. A: Ba B: Mn

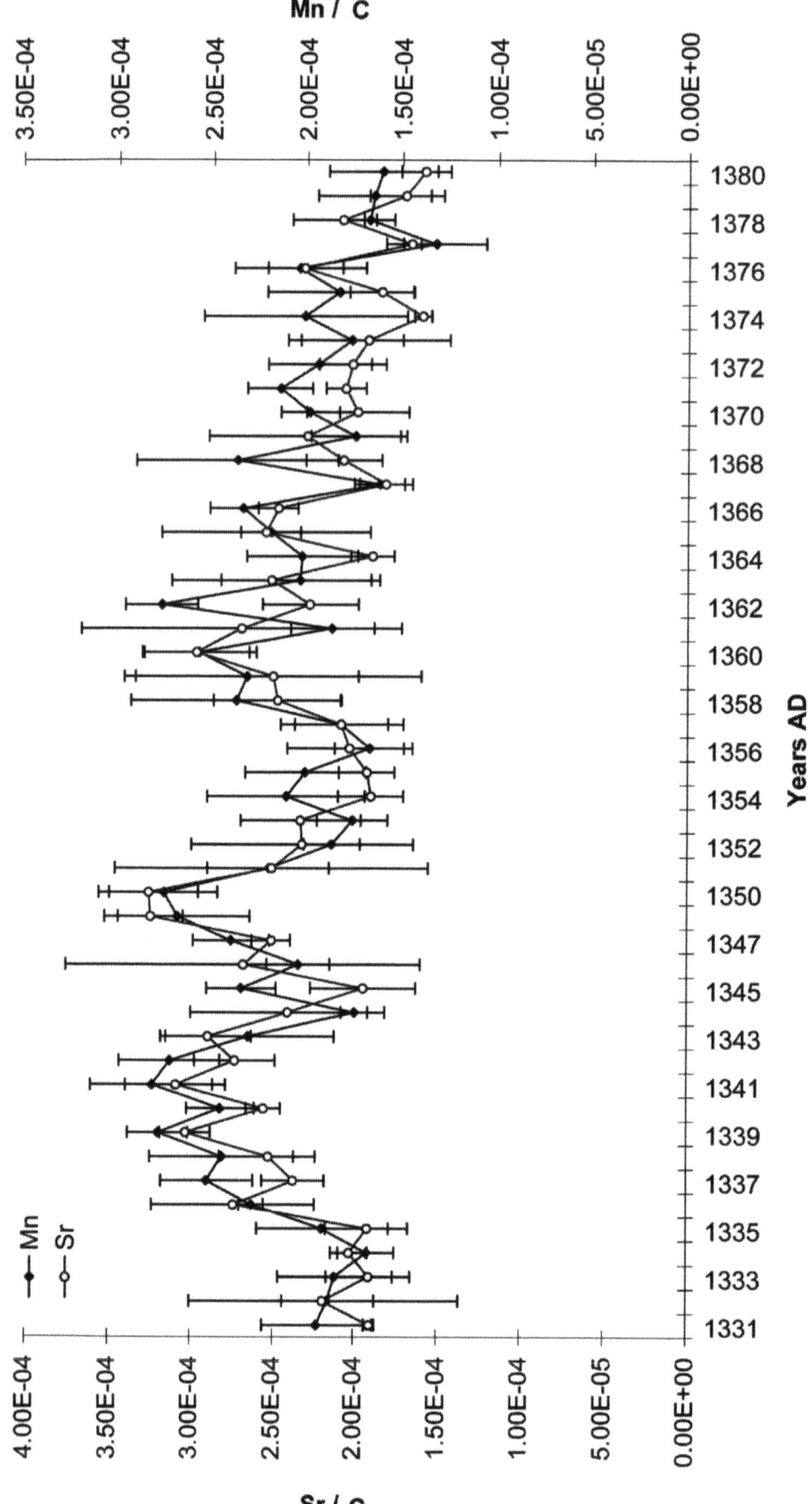

FIGURE 4.12: LA-ICP-MS analysis. Mn and Sr in *Agathis australis* 1331-1380 AD

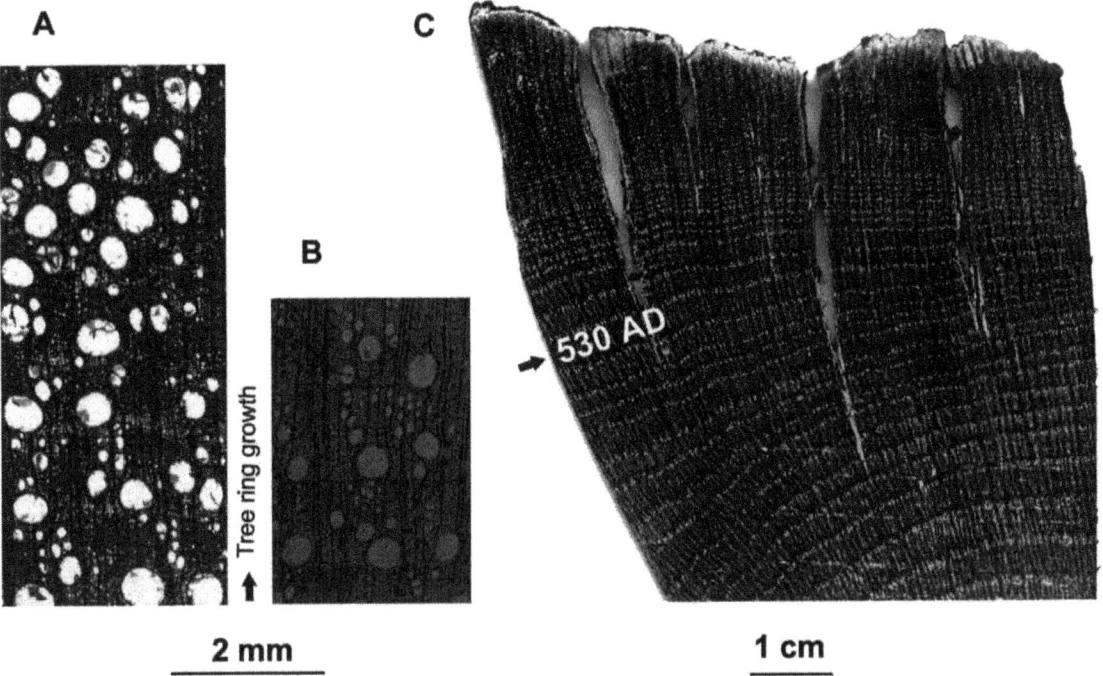

FIGURE 4.13: A: Light microscope slide (stained) showing the cellular deterioration of the tree rings from 534 onwards, the top area of the slide shows where conditions have become so extreme it is no longer possible to tell separate growth years apart. B: Light microscope slide (natural) showing normal growth rings. C: Sub-sample, sanded and chalked to show the 536 onset and deterioration

the given evidence of contamination, conclusions can only be tentative.

Figure 4.14 shows the agreement between the ratio patterns for Ca, Mn, Sr and Fe in the 25 year sequence analysed by LA-ICP-MS. The patterns for these elements match well and the ratios are similar suggesting a common contamination source. However it seems odd that the levels show so much variation if these patterns were entirely derived from saturation in the bog environment. Although impossible to prove given the obvious overlying impact of the burial environment, some element patterns did show possibly significant changes around 535/536 AD. The best examples of this can be seen in figure 4.15, where the results for Ba and La are plotted. Both these elements were found at similar ratios in the bog and terrestrial samples, so it is possible they are not derived from the bog environment. Ba shows increased heterogeneity from 530 to 537 AD and La is only present in the early wood of year 535 AD. It was noted that when the sample was analysed by solution ICP-MS, over thirty elements were detected, as opposed to the laser ablation analysis which only produced reliable data for nine or ten elements.

4.5.4 Outcomes

This pilot study was important as it began to explore the issue of the impact of the chemistry of the environment of preservation on older wood samples. This issue will be considered in further detail, and in a different context, in Chapter 6.1. Given that the results showed that there had clearly been some effect on the chemistry of the tree rings from the bog environment, it was decided that wood samples from such an environment could not be reliably used to trace changes in environmental chemistry in the years of tree ring formation. Preparation of samples for analysis by solution ICP-MS contributed to the development of the methodology in terms of experimentation to determine the optimum ashing time for this wood type. An opportunity was also provided to directly compare the range of elements detected in the same wood type by the two different sam-

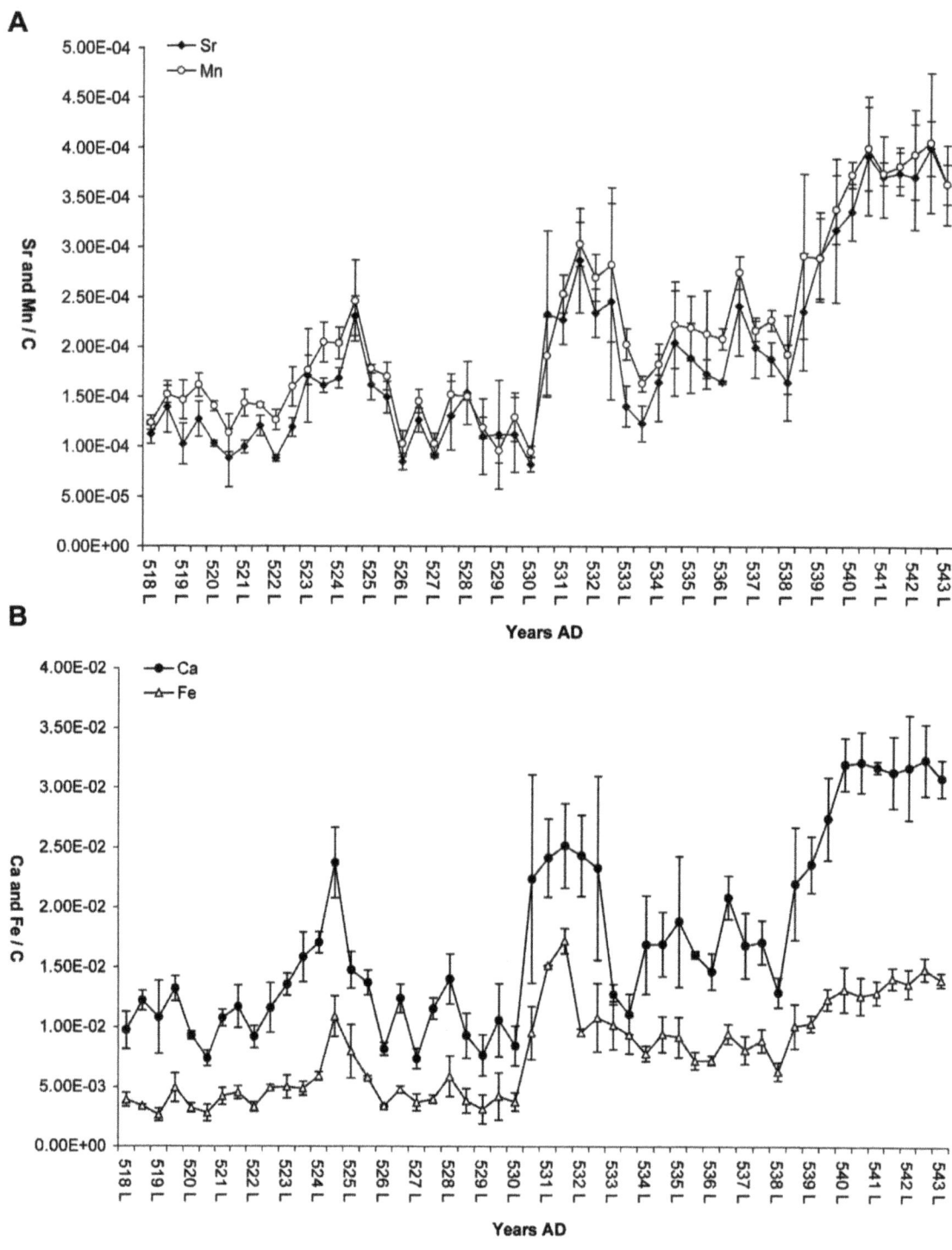

FIGURE 4.14: LA-ICP-MS analysis. A: Sr and Mn in early and late wood 518 - 543 AD. B: Ca and Fe in early and late wood 518 - 543 AD. (Only late wood is labelled on the graph)

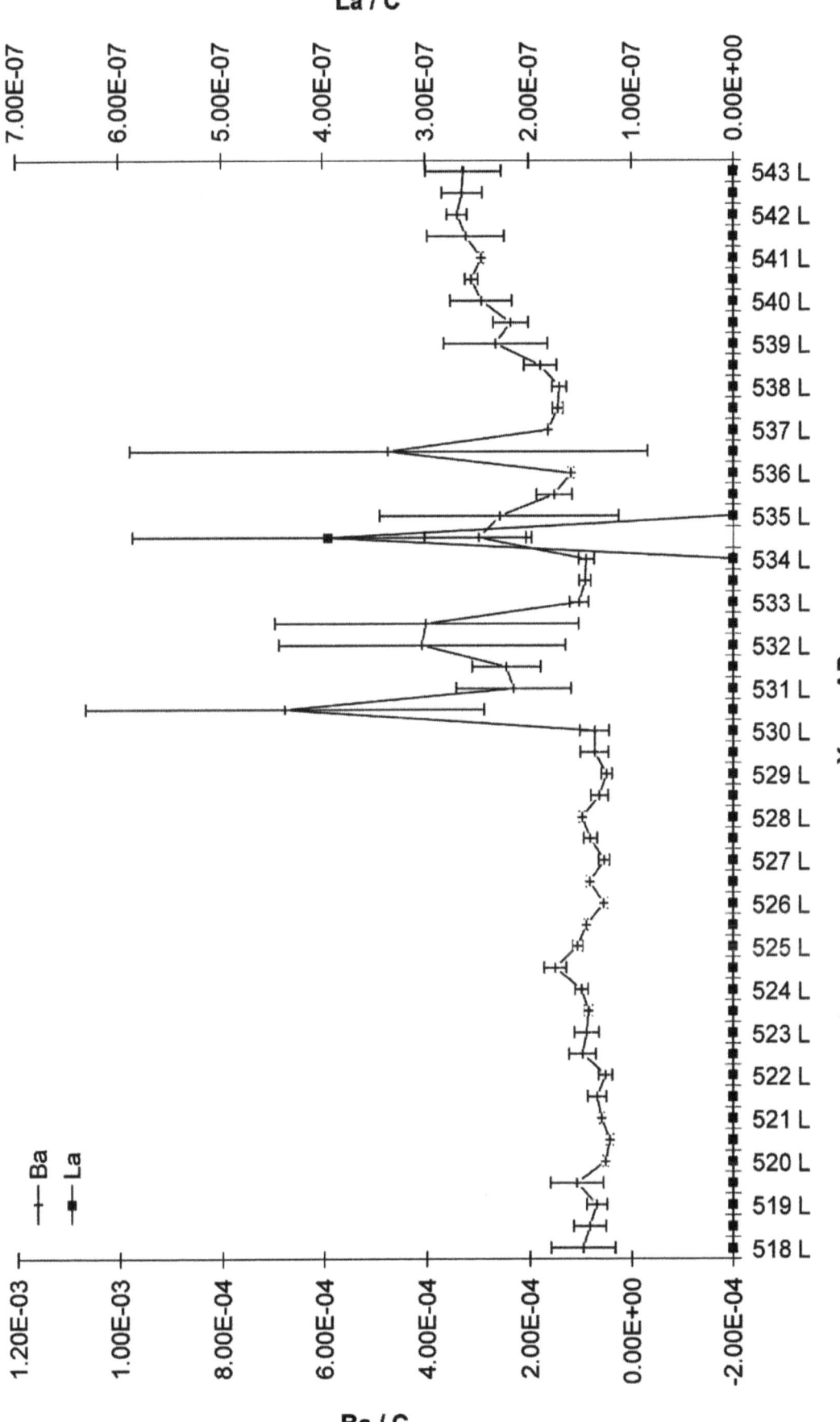

FIGURE 4.15: LA-ICP-MS analysis. Ba and La in early and late wood 518 - 543 AD. (Only late wood is labelled on the graph)

TABLE 4.1: Concentrations (ppm) of selected elements in bog and terrestrially grown *Quercus robur*.

Element	Bog *Quercus*	Terrestrial *Quercus*
Be	530	85
Al	2600	130
Ca	0.17	0.14
V	17	2.8
Mn	110	16
Se	4.5	2.9
Sr	0.0026	0.0025
Y	0.00073	0.00033
Zr	0.015	0.0039
Sn	0.0019	0.0017
Ba	0.0024	0.0018
Ce	0.0014	0.00077

ple induction techniques, (see Chapter 3, table 3.2 for further examples).

4.6 Pilot study 5

4.6.1 Introduction

Pilot Study 5 was the most in depth of the studies completed as it did not rely on absolutely dated wood samples obtained from chronologies, but involved a fieldwork element. With the aim of obtaining further samples from areas close to volcanoes with a known eruptive history, and to put into practice aspects of the developed sampling methodology (see section 3.1), contact was made with the Icelandic Forestry Commission. Two sampling sites in south-west Iceland were selected. The first site, in the Thjorsardalur valley, was a forestry plantation established in the 1960's. It was situated approximately eighteen kilometres due west of the Mt. Hekla volcano and, whilst receiving a major ash fall during the eruption of 1970, was not in the main fallout zone for the three subsequent eruptions (1980/81, 1991 and 2000). The second site, Tumastadir, was situated approximately thirty kilometres south-west of Hekla, to the east of the village of Hvolsvöllur. This site received tephra during the 2000 Hekla eruption and during the 1973 eruption of Heimey in the Westman Islands. The choice of species for analysis was limited to non-native trees, as the native dwarf-birch is unsuitable for dendrochemical analysis due to indistinct annual ring growth. The species selected for sampling was *Pinus contorta*, a species which has been used successfully in a dendrochronological and a dendrochemical context by Legge et al. [84].

Aims for analysis were to determine whether a chemical signature could be identified in any of the growth rings covering the various eruption years. It was also the aim to investigate any observed association of possible signatures with the direct deposition of tephra. Also, it was hoped to ascertain if, given the chemical background which might be supplied by the relatively immature volcanic substrate, it would be possible to trace any event at all.

4.6.2 Sampling and analysis

Samples were collected from Thjorsardalur (64°N, 19°W), at an altitude of 150 metres. A five millimeter diameter, Teflon coated, Swedish increment borer was used to extract cores from four trees at a height of one and a half metres from the ground surface. Four cores were taken from the western exposure of the trees, facing the dominant wind direction between the site and the volcano. An additional core was taken from the eastern exposure of tree three to provide the basis for an investigation of the chemical variability in a single tree. Cores were stored in plastic straws to prevent contamination. They were counted and prepared for sampling via LA-ICP-MS by being mounted in short sections on wooden plinths for ease of processing in the laser chamber.

Samples were collected from Tumastadir (64°N, 20°W), by Hrafn Oskarsson of the Icelandic Forestry Commission. Using a chain saw, radial discs (or cookies) were obtained from six trees at one and a half and three meters from the base. Samples were labelled and sealed in plastic bags for transportation to England. Following drying, they were prepared for analysis by LA-ICP-MS. Samples were counted and the sample surfaces were rigorously cleaned using a steel razor blade. They were then sub-sampled into strips and, as with the cores, mounted to fit the specifications of the sample chamber. Indistinct ring boundaries were identified and marked under a microscope to facilitate sampling accuracy in the laser ablation chamber. Several of the discs from Tumastadir developed fungal growth and could not be used for analysis. Two that were free from this contamination were selected for primary analysis. Tree 1 was sectioned into 5 radii (figure 4.16). Radii 1 consisted of a complete radial sequence (early and late wood) from pith to bark: 2001 - 1973. Four, ten year replicate sequences (1B, 1C, 1D, and 1E) were taken along this radii. Radii 2 was another complete radial sequence (early and late wood) from pith to bark: 2001 - 1973. Radii 3, 4 and 5 were ten year sequences along different radii of the same tree. Tree 2 was sectioned into three radii (2A, 2B and 2C) and these were analysed from 2000 back to 1983 for both early and late wood (figure 4.17 A). Five cores from Thjorsardalur were analysed from the pith (between 1970 and 1974) to the early wood of 2001. In the case of the two cores from tree 4, rings beyond 1998 were not analysed due to discolouration of the outer rings (possibly in response to some form of fungal growth). To investigate variation within the trunk, a comparative core was analysed from the eastern exposure of tree four. Figure 4.17 B shows a core being extracted at Thorsardalur, C shows a mounted core after analysis. Data were calibrated to ^{13}C and are presented as ratios to C.

4.6.3 Results summary

Of over 30 elements scanned for only 12 were reliably above detection. The results provided no elemental patterns which could be positively linked to eruption years. The only patterns to emerge were found in terms of the similar behaviour of specific elements in individual cores and radii. For example, an overall agreement between Mn, Sr, Ca, and Rb within individual cores and radii was noted in most of the samples. Figures 4.18 and 4.19 illustrate this for Mn and Ca in the four cores from Thjorsardalur. Whilst the patterns for Ca and Mn match in each separate core (as do the data for Sr and Rb), these elements did not match up from tree to tree, or within the same tree.

The association of groups of elements which are closely positioned on the periodic table is a naturally occurring phenomena. Sr, Ca and Rb are all alkaline metals commonly found in association with Mn (a litho-file). The fact that this association occurs is useful when checking for instrumental fluctuations. For example, in figure 4.19, part B, there was a break in ablation around 1982 and the sample chamber was opened and resealed. The differences between the patterns for the two elements after this point shows that the instrument required some adjustment.

Figure 4.20 shows a direct comparison between the Mn sequences from cores 1 and 4, plus the duplicate core from the opposite side of tree 4 (B). This is a typical example of how variable element sequences were within the same tree as well as between different trees growing only a few meters apart. Two hypotheses to explain this observed variability are, first, that the tree rings are so heterogeneous that one ablation in a ring does not provide a true representation of the chemistry of that ring. Or, second, that the underlying immature volcanic bedrock and successive ash layers provide such similar, consistent amounts of various elements to those expelled by various eruptions that no change in the chemistry is discernable. This being the case it is certainly possible that the chemistry of these tree rings represents background noise and that the within ring variation has been magnified by instrumental fluctuations and comparison with low levels of fluctuation between rings.

The samples from Tumstadir produced similarly inconclusive results. Due to instrumental failure only radii 3, 4 and 5 from tree 1, and 2A, 2B and 2C from tree 2 were reliably analysed. Of the data collected, no discern-

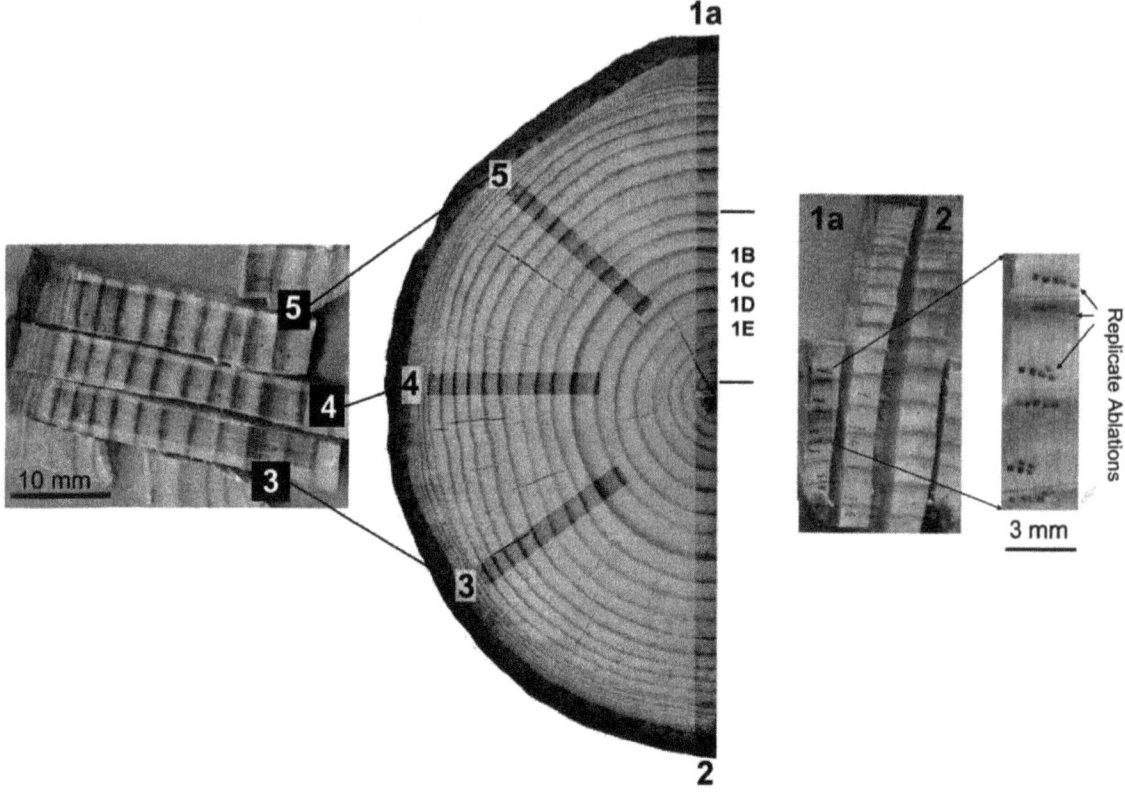

FIGURE 4.16: Analysed radii of Tree 1 from Tumstadir.

able chemical traces were found to link specific years with volcanic eruptions. Again, Mn, Ca, Sr, and Rb showed good agreement with one another, as did Al and Ni (also closely associated elements). Examples of the type of data produced are given in figures 4.21 and 4.22. These samples appeared marginally more homogenous than the Thorsardalur cores, although with the same lack of apparent agreement between the ratio values for the same tree ring along a slightly different radius. The overlapping error bars are indicative of essentially the same ratio value for the sequences and it could be argued that what is being observed is largely background noise against which any observable fluctuations are too small to be seen.

From these preliminary results many issues were raised, with more questions than answers. These include: (i) Is *Pinus contorta* too heterogeneous a wood type for analysis by LA-ICP-MS? (ii) Is the LA-ICP-MS technique in need of further work to improve the 'instrumental error' and thus duplicability of data sets? (iii) Is it likely that potential chemical tracers of volcanic eruptions are not being detected? (iv) Is it possible that the relatively short Icelandic growth period did not coincide with the main volcanogenic changes in environmental chemistry, and that no trace was taken up due to the inactivity of the trees?[2]. It was concluded that it appeared unlikely that a volcanic signature may be found in tree rings from such a growth environment via LA-ICP-MS. However the potential for a signature may exist if the tree rings were analysed by a technique which detected a wider range of elements. Nearly all the elements detected via LA-ICP-MS were essential elements, the distribution of which may have been preferentially controlled by the tree itself rather than by external determinants. If other less common elements were detected it might be possible to find a volcanic trace.

---

[2]Studies such as [86] do suggest however, that certain elements will be taken up through the bark whether the cambium is active or not.

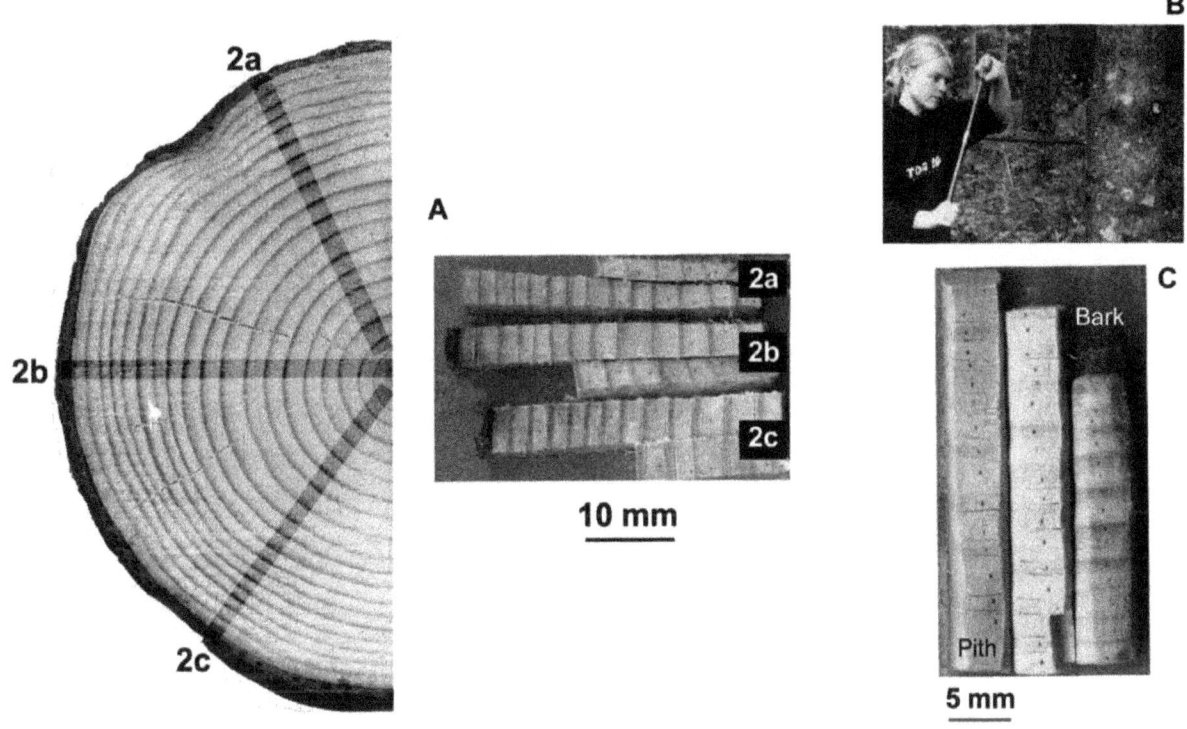

FIGURE 4.17: A: Analysed radii of Tree 2 from Tumstadir. B: Core extraction at Thorsardalur C: Example core mounted for ablation

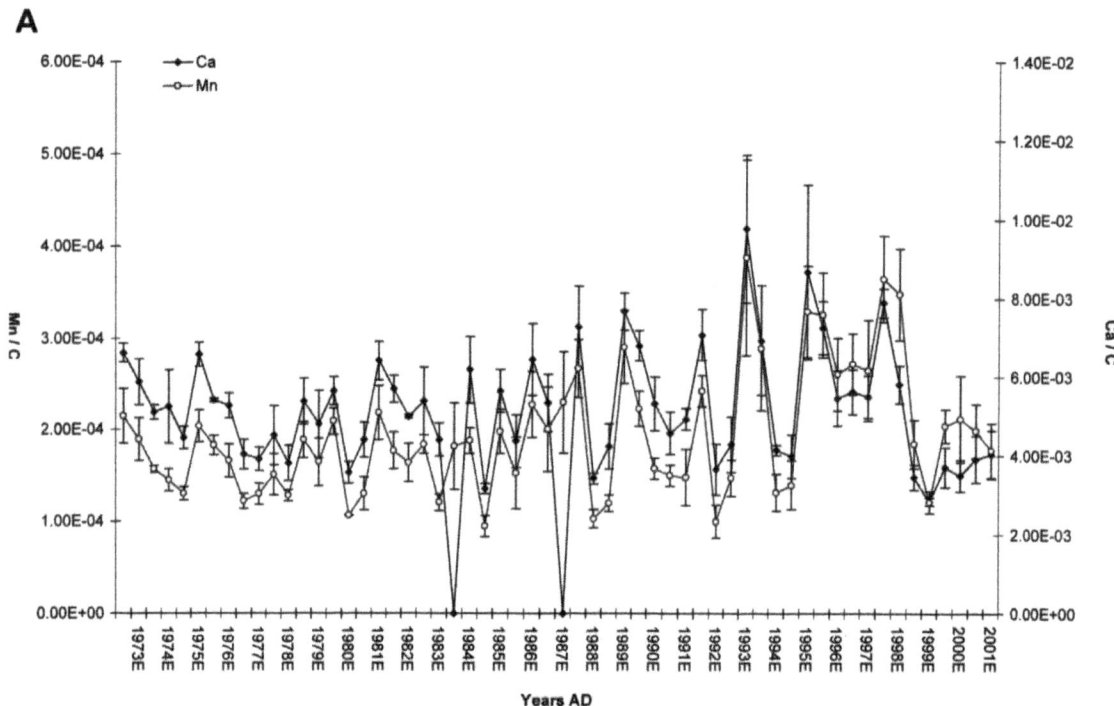

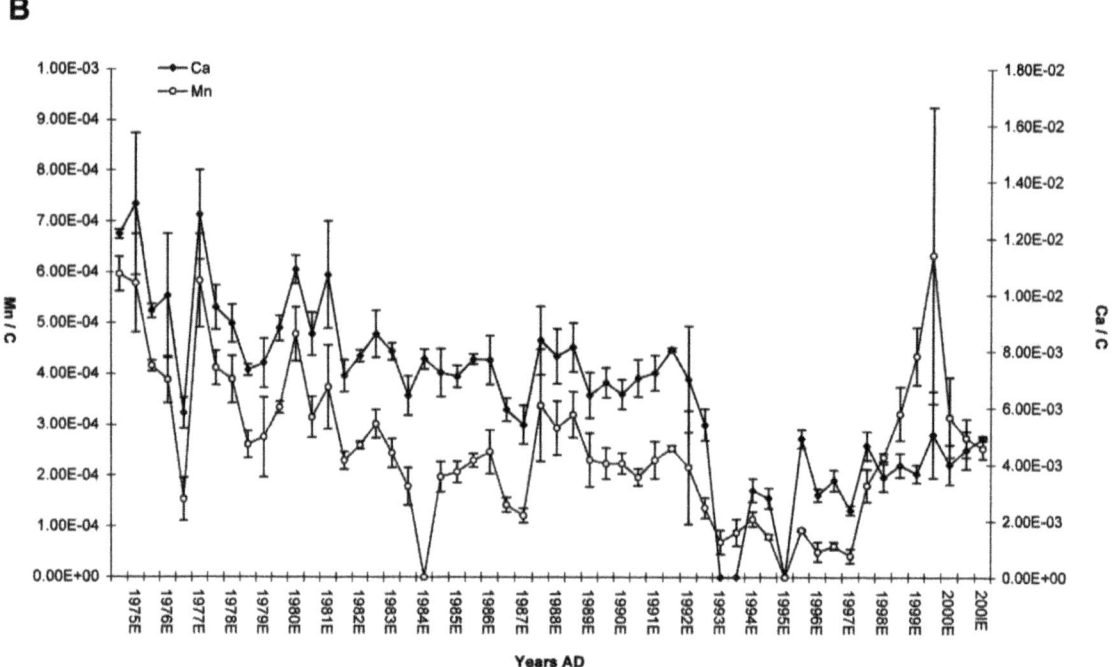

FIGURE 4.18: LA-ICP-MS analysis. Ca and Mn early and late wood sequences from A: Core 1, tree 1 Thorsardalur (1972 early wood to 2001 early wood, early wood only labelled). B: Core 2, tree 2 Thorsardalur (1974 late wood to 2001 early wood, early wood only labelled).

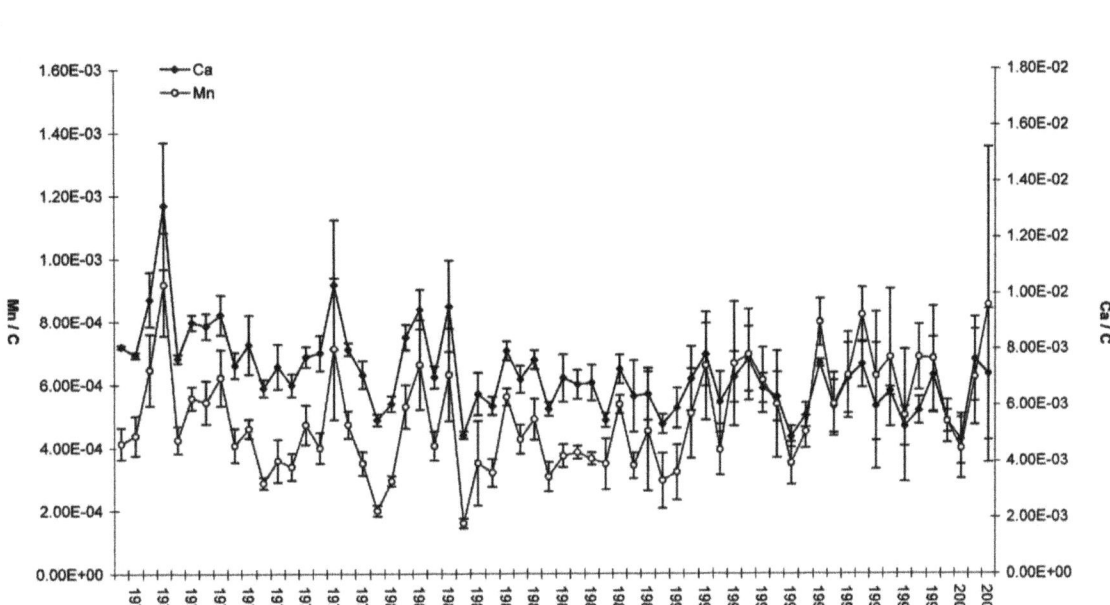

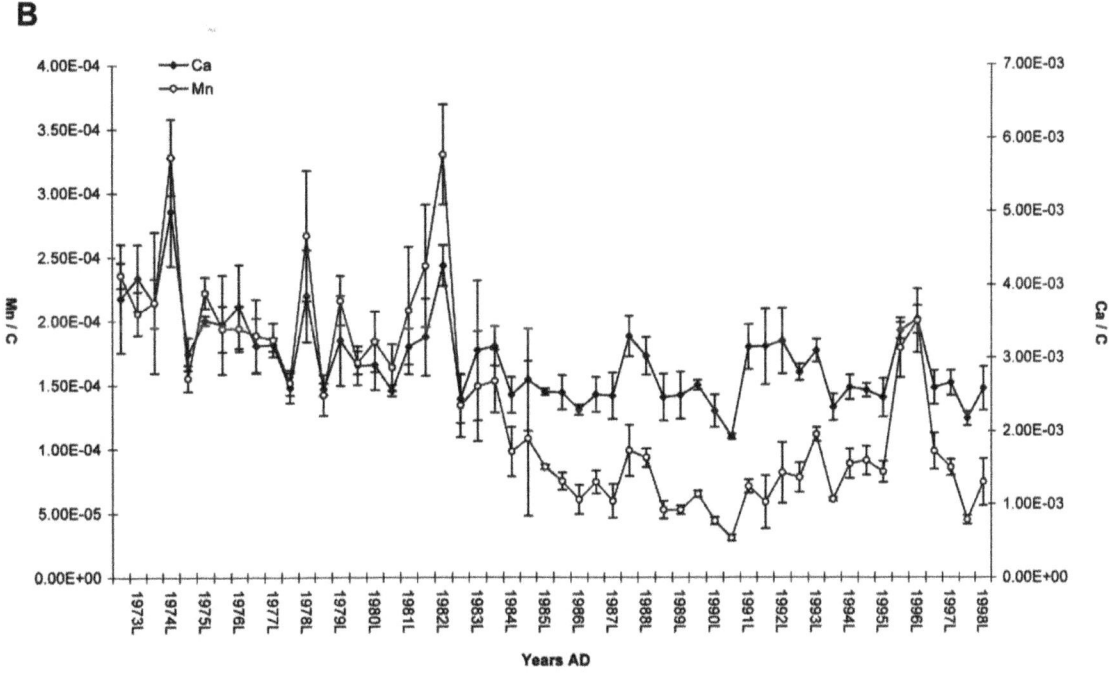

FIGURE 4.19: LA-ICP-MS analysis. Ca and Mn early and late wood sequences from A: Core 3, tree 3 Thorsardalur (1970 late wood to 2001 early wood, early only wood labelled). B: Core 2, tree 2 Thorsardalur (1973 early wood to 1998 late wood, late wood only labelled).

CHAPTER 4: PILOT STUDIES

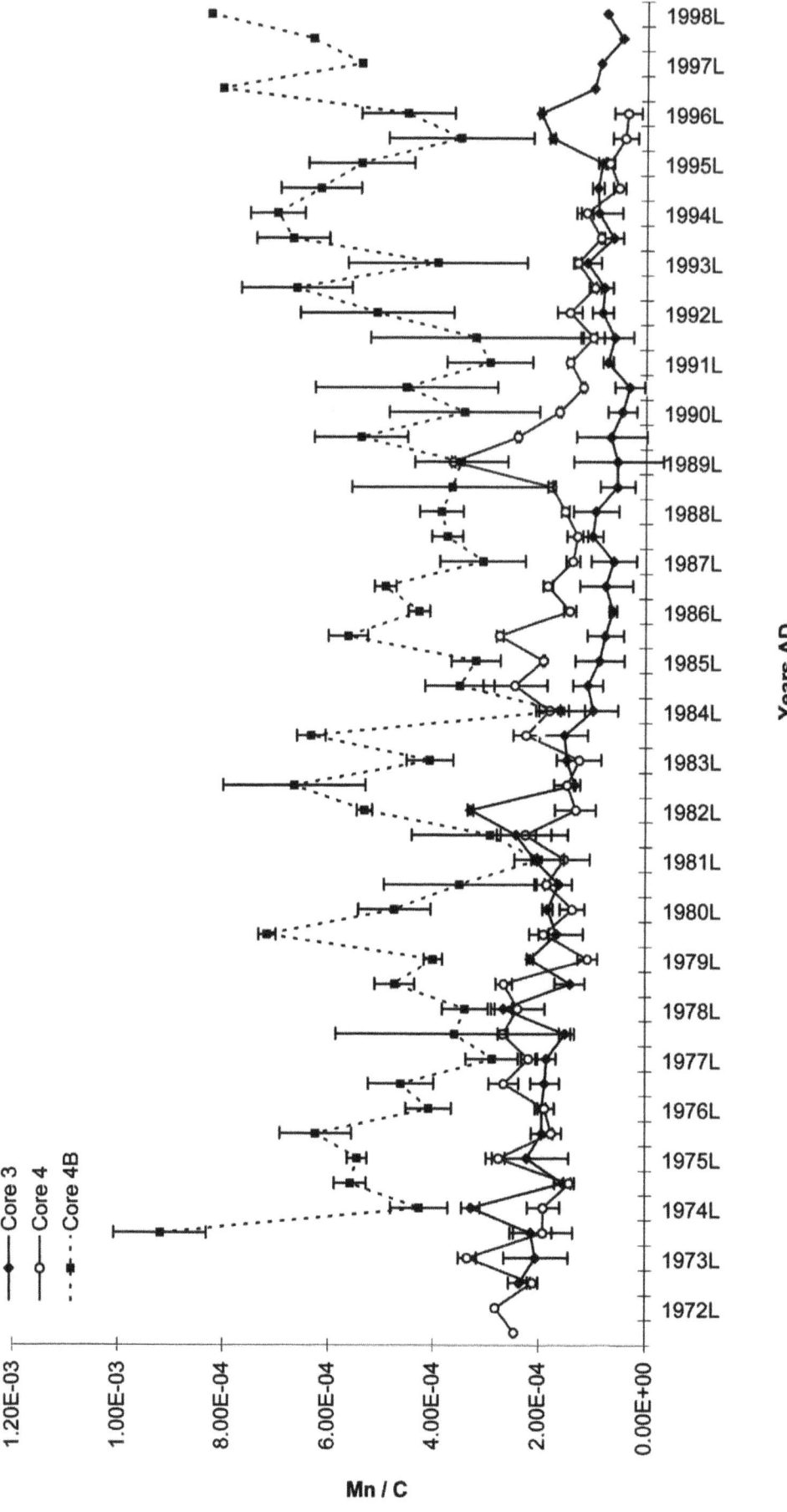

FIGURE 4.20: LA-ICP-MS analysis. Mn in early and late wood sequences from Cores 1, 4 and 4B - the replicate core from the opposite side of the trunk of tree 4, Thorsardalur (1972 early wood to 1998 late wood).

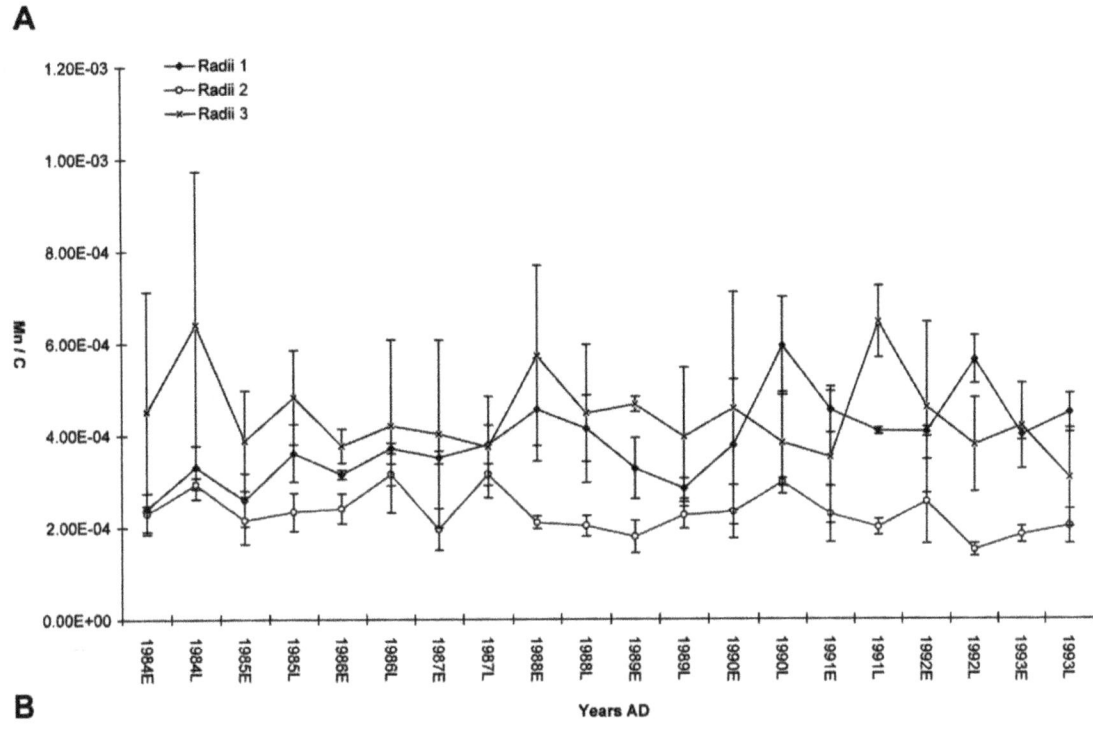
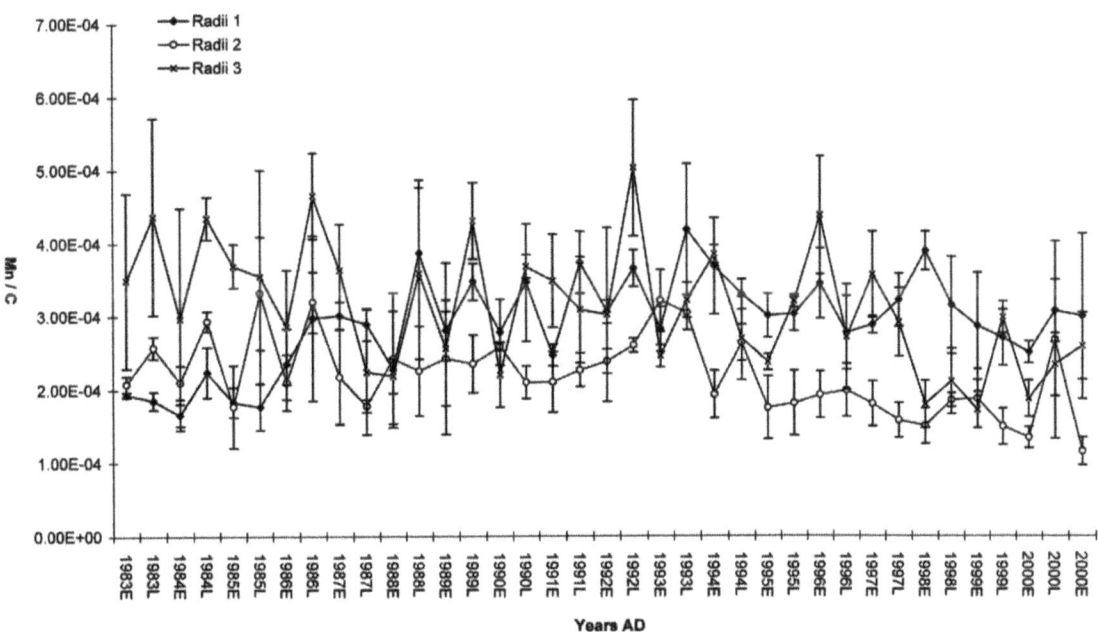

FIGURE 4.21: LA-ICP-MS analysis. A: Mn in Radii 3, 4 and 5 from Tree 1, Tumstadir (1984 early wood to 1993 late wood). B: Mn in radii 2A, 2B and 2C from Tree 2, Tumstadir (1983 early wood to 2000 early wood).

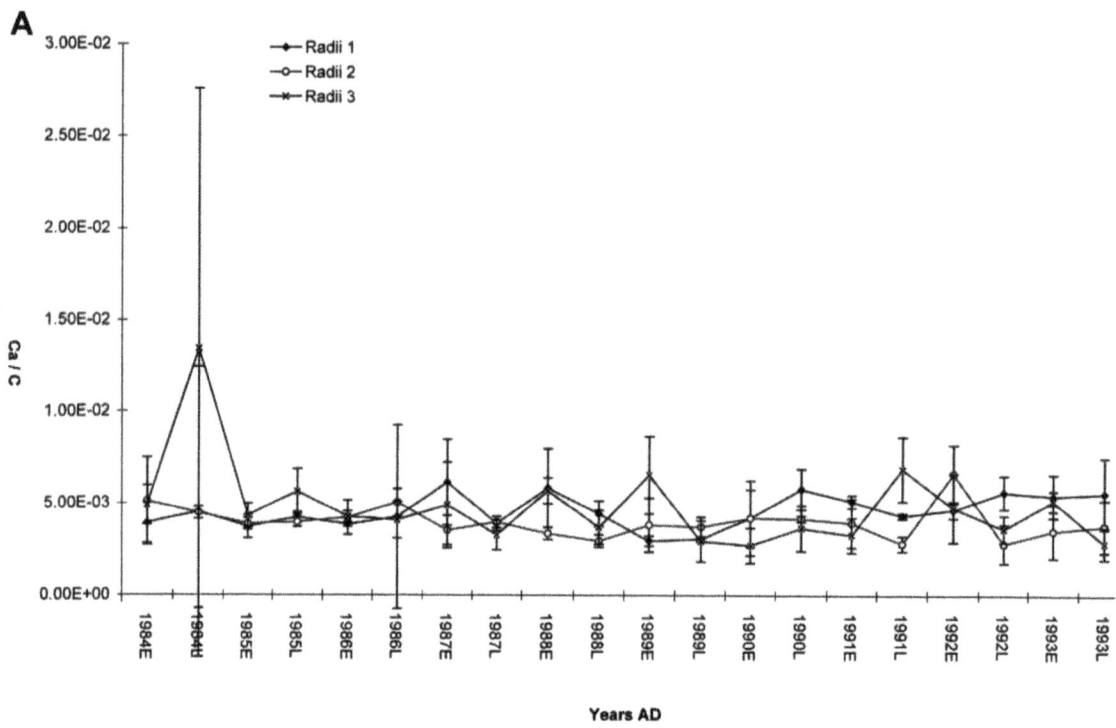
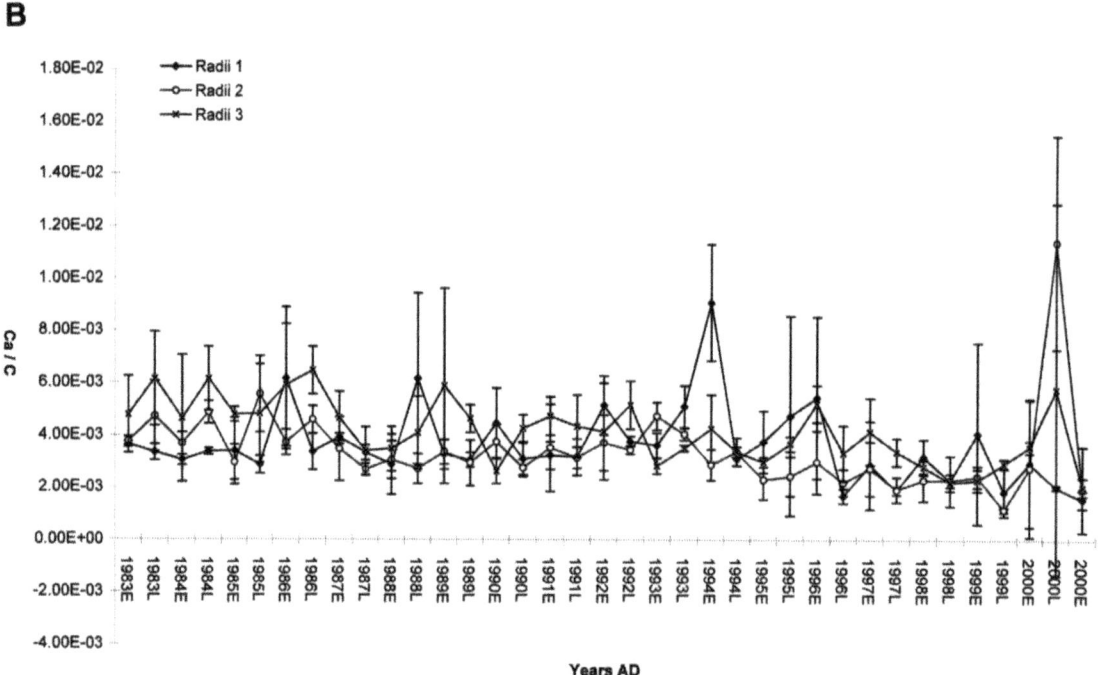

FIGURE 4.22: LA-ICP-MS analysis. A: Ca in Radii 3, 4 and 5 from Tree 1, Tumstadir (1984 early wood to 1993 late wood). B: Ca in radii 2A, 2B and 2C from Tree 2, Tumstadir (1983 early wood to 2000 early wood).

4.6.4 Outcomes

The results and processes involved in the study were very useful in terms of future site selection and development of the procedural and analytical technique (the types of issues discussed in sections 3.1 and 3.3. In addition, experience in the field contributed greatly to the development of sensible sampling and storage strategies. For example, problems were experienced with mould growth during the transportation of the samples back to the lab. This meant that certain years had to be left un-sampled, resulting in shorter, less appropriate sequences. For all subsequent sampling, plastic straws and bags were aerated in order to encourage air circulation and quick drying.

Along with the previous four pilot studies, the results raised real questions as to the overall use of LA-ICP-MS for the analysis of tree rings to find volcanic signatures. This major pilot study was a final attempt to use LA-ICP-MS to explore this main research objective. The data, as found previously, brought to light the key problems of sequence replication and detection of an insufficient number of elements. It also highlighted the problems with joining data sets from separate days of analysis, illustrated in Chapter 3 (figure 3.2). From this study it was finally concluded that solution ICP-MS would be a preferable mode of sample induction, as not only had it been shown in previous studies to detect a far wider range of elements, but in preparing solutions of tree rings any inhomogeneity would be averaged out.

4.7 Pilot study 6

4.7.1 Introduction

Requests were made to various overseas institutions for wood samples suitable for trial analysis by solutions ICP-MS. A piece of *Picea glauca* was obtained from the Tree-ring lab at the Lamont-Doherty Earth Observatory, Columbia University. The sample was a quarter radial section from a tree grown in Alaska. The main aims for analysis were to prepare a long sequence of samples in solution, in order to develop a successful methodology for solutions preparation and analysis by ICP-MS (see section 3.4). The particular sequence selected covered a time period when there were no major eruptions, in order to investigate the type of background elemental patterns found in this tree species.

4.7.2 Sampling and analysis

The radial section was sub-sampled into 1.5cm thick slices and each slice was cleaned by removal of the outer surface via a combination of sanding and slicing with a steel blade (see figure 4.23 for an illustration).

150 samples (from 1600 to 1750 AD) were dissected. Samples of waste material generated during the cleaning process were wet and dry ashed for various periods of time, and at various temperatures as a precursor to the more controlled experimental design process described in section 3.4.2. It was found that 0.2g of sample, dry ashed in a muffle furnace at 375°C for 3.5 hours produced the most effective dissolution in concentrated nitric acid, in the shortest possible time (although some filtration was necessary). Samples were made up to 20mls with ultrapure deionised water.

4.7.3 Results summary

The lower limit of detection was relatively high due to high procedural blanks. However, elemental concentration patterns were successfully detected for lithium (Li), Al, Ca, Ti, V, chromium (Cr), Mn, Fe, Co, Ni, Cu, Zn, As, Rb, Sr, Y, Zr, Rh, Ag, Cd, Sn, Sb, Cs, Ba, La, Ce, praseodymium (Pr), Nd, Sm, europium (Eu), Gd, terbium (Tb), dysprosium (Dy), holmium (Ho), erbium (Er), Lu, gold (Au), Pb, Th and U. The end part of the data set (1737 to 1750) was lost due to instrumental drift, however for the majority of the sequence the instrument was stable and the error was low. The elements detected showed a wide variety of patterns, some examples of which are given in figures 4.24 to 4.26.

The widely contrasting elemental patterns displayed in the data could be due to any number of external influences on tree ring chemistry. The well defined rise in Cs concentrations around 1680 and the peak of Pb in the

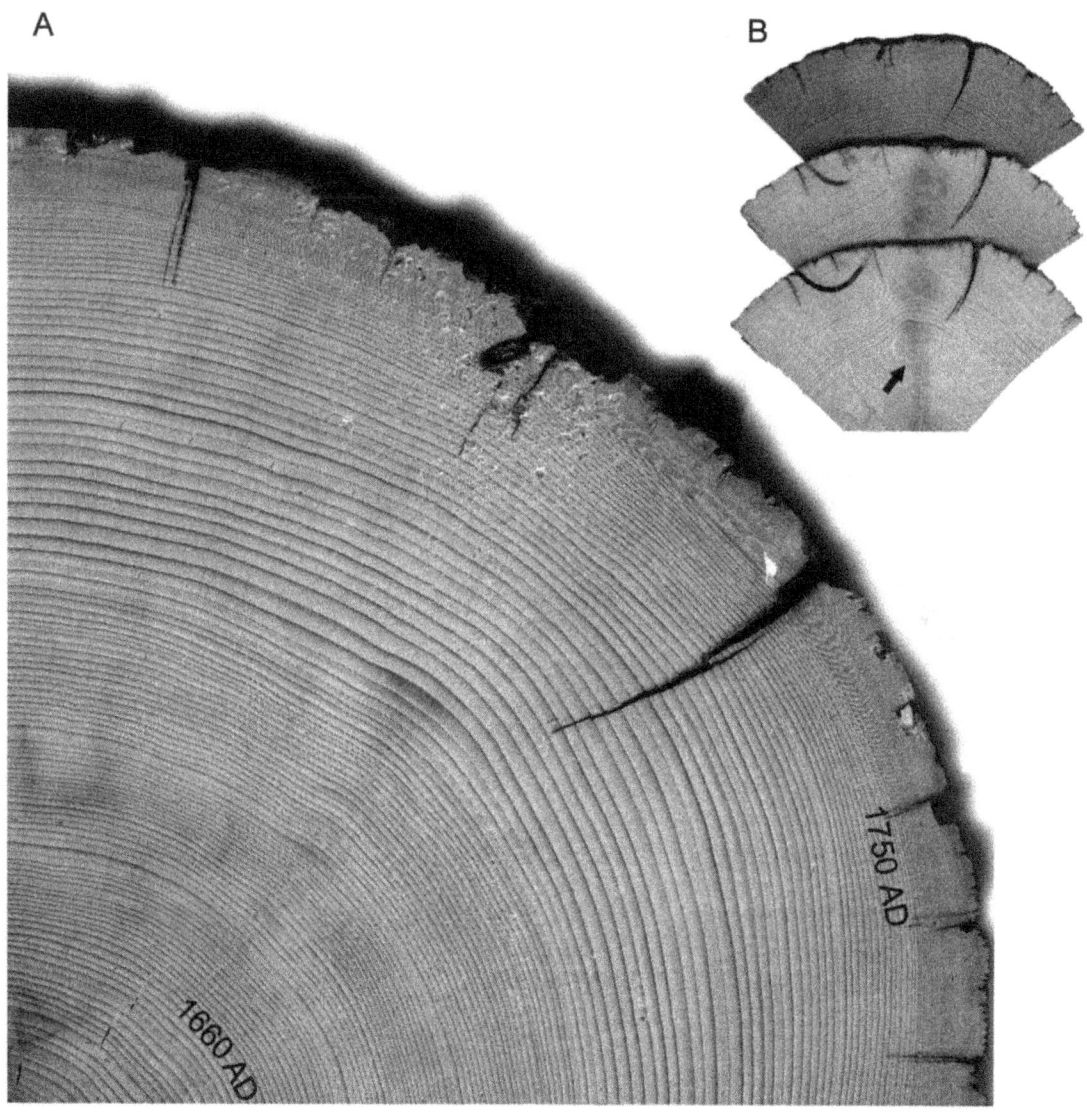

FIGURE 4.23: A: Main sample. B: Steps in the cleaning process, increasingly clean samples. Arrow highlights a branch scar which was avoided during dissection.

CHAPTER 4: PILOT STUDIES

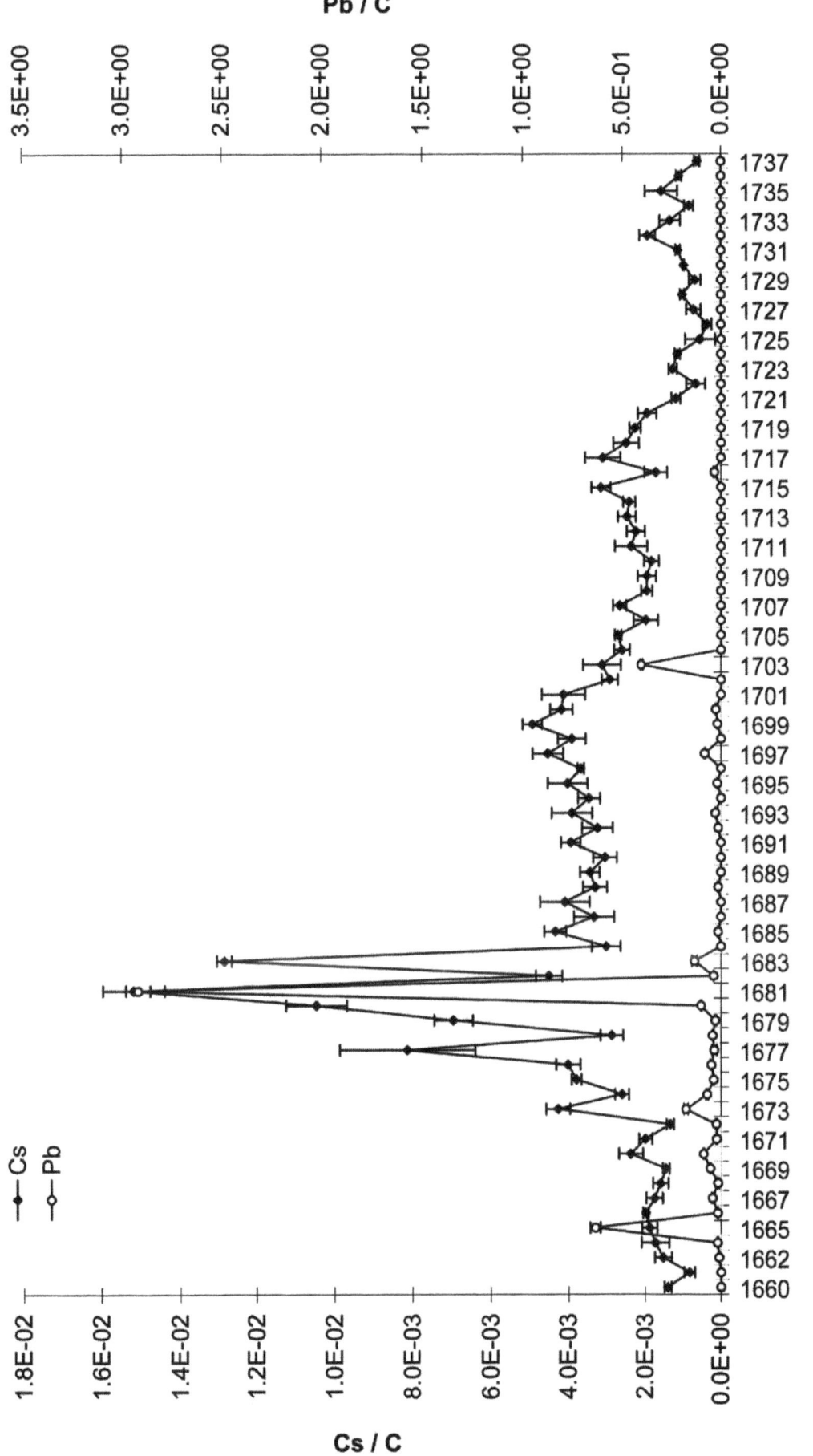

FIGURE 4.24: Solution ICP-MS analysis. Cs and Pb in *Picea glauca* (1660-1737)

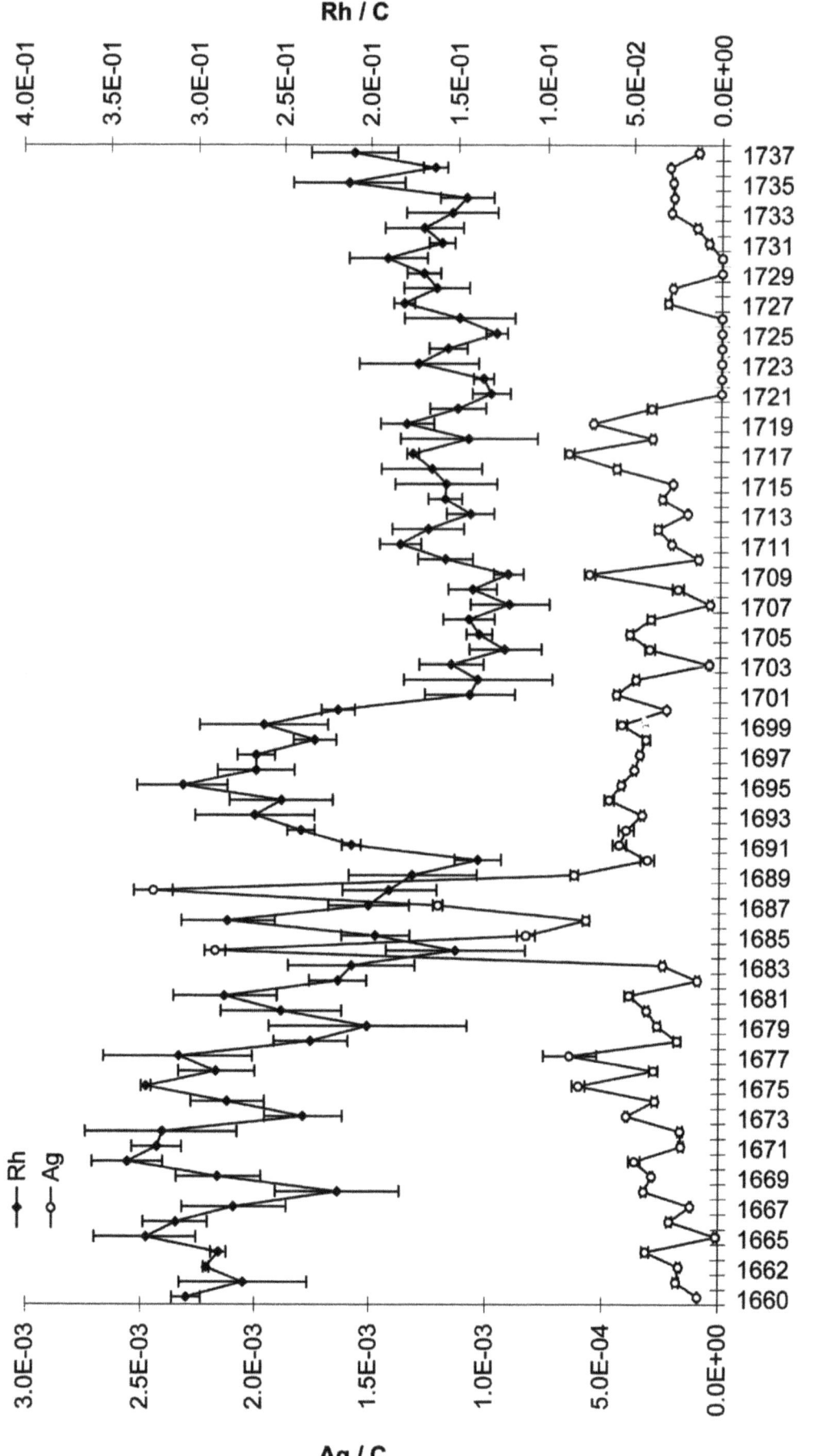

FIGURE 4.25: Solution ICP-MS analysis. Rh and Ag in *Picea glauca* (1660-1737)

CHAPTER 4: PILOT STUDIES

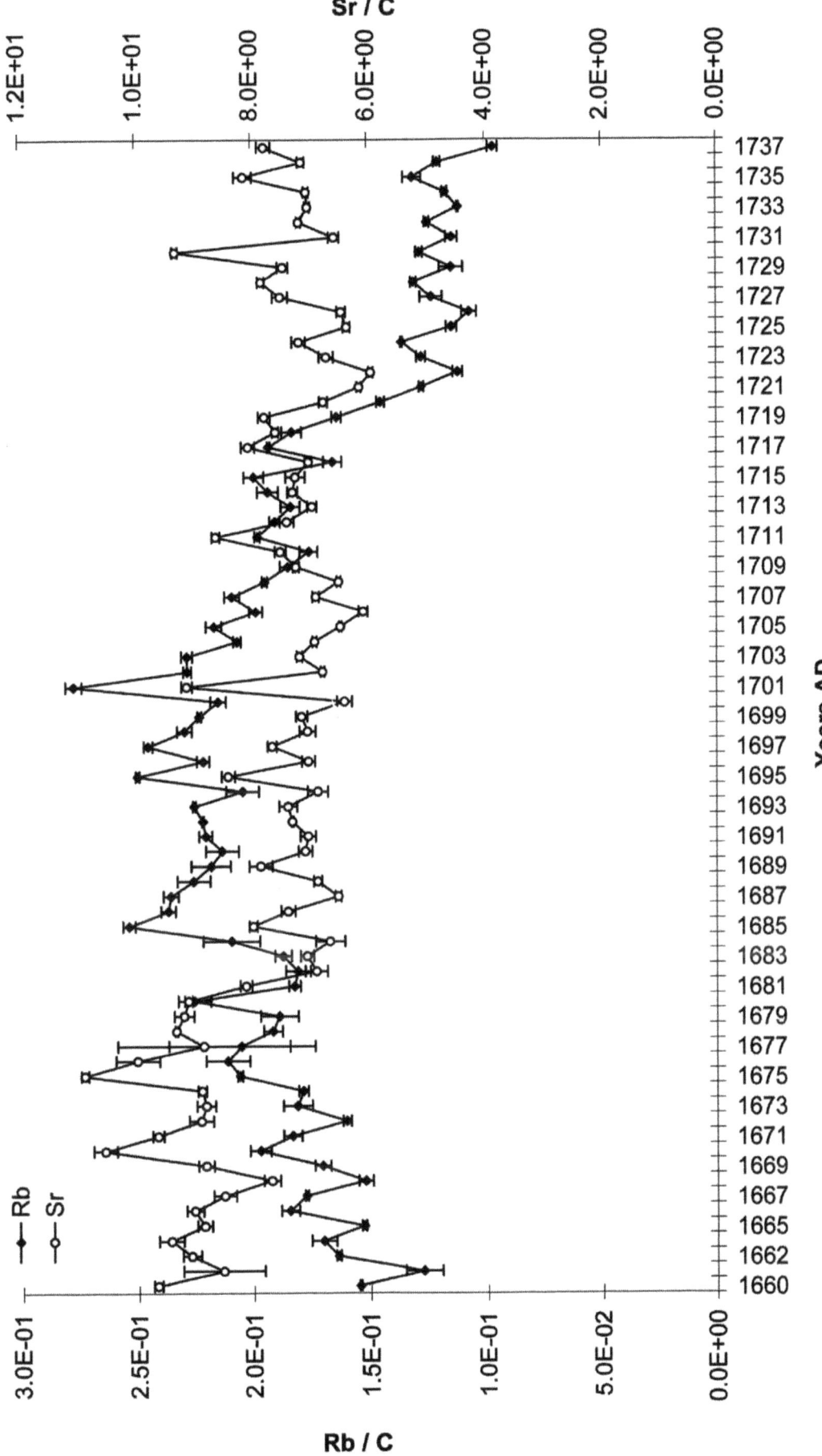

FIGURE 4.26: Solution ICP-MS analysis. Rb and Sr in *Picea glauca* (1660-1737)

same year are intriguing, as are the Ag peaks in 1684 and 1688 and the rise in Rh from 1691 to 1700. Although no major eruptions occurred through this period there were some over VEI[3] 5, these were in 1663, 1667, 1673, 1680, 1707, 1721 and 1739; it could be argued that some of these dates correspond with depletions of Rb and Sr, and that Cs, Pb and Ag are responding to the eruption in 1680. However there is nothing more than speculation to link these events. In all likelihood, the volcanic events in these years were not sufficiently large or close enough to the tree in question to be detected at all, and any possible correlation is most likely to be entirely coincidental. The patterns observed could equally be due to any number of other possible influences, not only from exogenous sources, but also from within the tree and as introduced contamination during the sample preparation procedure. No conclusions could be drawn from the data in terms of the observed changes in chemistry, however the successful detection of a wide range of elements following the procedure of dry ashing and dissolution was highly significant in terms of the methodological development of the project.

4.7.4 Outcomes

The aims and outcomes of this study were mainly in terms of methodological development, specifically, that it was found that solution induction ICP-MS could successfully be used to analyse a long sequence of individual tree rings, with low error, for a wide range of elements, including the rare earth elements. This overall result was useful in shaping the methodological development and critically, in showing the advantages of a solution induction method for ICP-MS over laser ablation in terms of elemental detection. The results of the analysis also served to highlight the care needed with interpretation of data sequences covering known, major volcanic eruptions. In this sequence strong patterns seemed to be showing exactly the type of sudden short term responses in the tree ring chemistry that might be expected in relation to volcanic eruptions. However, it is unlikely that any of the volcanic eruptions during this period would have had any effect on tree ring chemistry. If such patterns can occur in tree ring sequences irrespective of volcanic eruptions, as seems likely, then the only way in which an association could ever be truly established would be via an actual chemical finger print for a specific eruption being found in a set of tree rings. As the only way to achieve this would be by detecting the maximum range of elements in a given sample, the logical way to proceed was analysis via solution induction ICP-MS.

4.8 Conclusions

These example pilot studies contributed to the shaping of the methodology, and to key steps in the decision making process which concluded with LA-ICP-MS, in spite of its advantages, being superseded by the use of solution ICP-MS (and ICP-AES) for further analysis. With an optimised strategy for site selection, sampling, sample preparation and analysis in place, the remainder of the analytical work was carried out in the form of two case studies. The next two chapters describe and discuss these studies, which consider the implementation of all of the above on modern wood samples (Chapter 5) and the complications of putting the same methodologies into practice in the context of archaeological samples (Chapter 6.1).

---

[3]The Volcanic Explosivity Index (VEI) is a scale developed to describe the explosive magnitude of a particular volcanic eruption. For a full explanation see Newhall and Self [104]

Chapter 5

Case Study 1 - Sweden

5.1 Introduction

A site was chosen to meet the specifications of the optimized sampling strategy (section 3.1). Samples were collected, prepared, and analysed by solution ICP-MS. A thorough investigation was made into the chemical variations within individual tree rings and at different heights in the trunk, also into the influence of site specifics such as soil chemistry and bedrock. This was so that analysis of the elemental data generated for the sequences of tree rings could include consideration of endogenous physiological and site specific factors in addition to attempting identification of exogenous environmental signatures.

5.1.1 Aims and objectives

The main aim of case study 1 was to conduct an holistic study into the tree ring chemistry of a specific wood type, utilising the optimised methodologies developed in chapter 3 for all stages of site selection, field sampling, wood storage, sample preparation and analysis. Specific criteria for this case study were:

1. To quantify the effectiveness of all steps of the designed practical procedure.

2. To measure variations in the chemistry of one ring at one height and at a range of different heights in one tree, with the aim of understanding aspects of physiological control on elemental distribution within a single tree ring.

3. To explore the various possible local exogenous contributions to the tree ring chemistry and growth pattern.

4. To determine whether wider ranging exogenous inputs such as anthropogenic pollution and volcanism could be detected within the tree rings.

5.2 Site selection

For the purposes of the study a site was required with shallow, mature soils and minimal exposure to anthropogenic pollution. It was anticipated this combination of factors would promote low background levels of various elements in the tree rings, and increase the probability that uptake paths through the bark and leaves would be dominant. A detailed record of the history of the site, and availability of site specifics such as temperature and precipitation were other criteria. With these aims in mind, the Siljansfors Försökspark (Siljansfors Experimental Forest), eighteen kilometres south west of Mora in south, central Sweden was selected. The forest lies between 60°52' N 14°19' W and 60°55' N 14°25' W, at an elevation of 210 - 425 metres above sea level. The deposition chemistry of the area is considered low level in terms of sulphur and nitrogen and has no marine influence. The vegetation is dominated by Scots pine (*Pinus sylvestris*), interspersed with Norway spruce (*Picea abies*), with some birch (*Be-*

tula sp.) on the lower slopes. The field layer consists mainly of dwarf shrubs such as blueberry (*Vaccinium myrtillus*), cowberry (*Vaccinium vitis-idaea*) and heather (*Calluna vulgaris*). The surface layer comprises a sheet of mesic mosses (eg. *Pleurozium schreberi* and *Hylocomium splendens*). The main growing season is approximately May - September. The soil is a shallow, cold climate podzol in rocky, well drained sandy soil. It has a greyish eluvial horizon above a reddish illuvial horizon mainly composed of organic matter with varying amounts of iron and aluminium. This is formed on a locally derived glacial till. The underlying bedrock is a Pre-Cambrian welded ignimbrite tuff - (Slirig porphyry: lower Dala series).

5.3 Field sampling

In the field, four trees (*Pinus sylvestris*) were sampled for possible analysis. Samples Tree A and Tree B were from fallen trunks and later disregarded for chemical analysis due to discolouration of the sapwood and evidence of partial decay. Samples Tree C and Tree D were selected for chemical analysis. Radial samples were extracted by Christer Karleson (see figure 5.1) of the Swedish University of Agricultural Sciences, Siljanfors. Samples were double wrapped in aerated plastic bags. Due to sub-zero temperatures at the time of sampling, samples remained in a semi-frozen state from sampling until arrival in the lab. In other situations samples would have been frozen as soon as possible after sampling to limit sap migration. The main purpose of sampling Tree C was to obtain a several hundred year record for analysis from a specially selected, newly felled tree. The tree was sampled at different levels in order to explore differences in the chemistry of the same tree rings at different heights in the tree. Tree D, from an almost identical site close to Tree C, was sampled primarily to provide a comparison sequence. However, it was also sampled because the tree in question (tree 76 in plot 102.22), had been part of an experiment to determine the effects of fertilisation on growth. Park records note that it was fertilised in May 1967, when it received ammonium nitrate, calcium and nitrogen (the actual mix used was Kalkammon salpeter: 4% calcium, 2% magnesium, 27.6% ammonium nitrate mixed with dolomite). This was applied manually as pellets at intervals on a regular 10.5 m grid, spread evenly over the area. It was anticipated that this fertilisation might be detectable in the growth ring pattern and / or some sort of elemental response. Soil samples were taken for each site in order to gain a better understanding of the growth environment and the relationship between soil chemistry and tree ring chemistry at each site. The field sampling information for Trees C and D is given in the following sections.

5.3.1 Tree C

Tree C, (60°53' 12.7" N 14°19' 51.3" E) was growing on the crest of a small hill, on gently sloping ground. The wider topography consisted of a glaciated landscape of rolling hills, valleys and lakes. The underlying soil was shallow and stony, including large (<15cm) angular clasts of the igneous bed rock which outcropped in several places nearby. The trunk was straight and evenly proportioned showing no sign of tension or compression wood. The tree was felled, and radial samples (c. 24cm thick) were taken at 20cm, 1.5m, 2.5m and 3.5m from the base of the trunk. A soil profile was accessed 1.5m from the base of the trunk and four horizons with undulating boundaries were identified. Soil samples of c. 5g were taken from each horizon from three positions at 1.5m around the base of the remaining stump. Table 5.1 provides a detailed description of each of the identified soil horizons.

5.3.2 Tree D

Tree D, (60°53' 14.4" N 14°19' 58.6" E) was growing in a very similar location to Tree C. In this case however the tree was on a slightly steeper slope on the side of a low, rolling hill. The underlying soil was very shallow and slightly stonier than for Tree C, including large (<15cm) angular clasts of the igneous bed rock. As before, a trunk was selected which was straight and evenly proportioned showing no sign of tension or compression wood. The tree was felled, and radial samples (c. 30cm thick) were taken at 20cm and 1.5m up the trunk. A soil profile was cleared 1.5m away from the base of the trunk and three horizons with undulating

FIGURE 5.1: A: Christer Karleson sampling Tree C. B: Sampling sections at various heights

TABLE 5.1: A detailed description of the soil horizons at the sampling location for Tree C

Depth (cm)	Colour	Description
0-5	Brownish black 2/1	Humic matter with plant material and occasional sub-rounded clasts of the bedrock 97% Organic 65% Discernable plant material (lichen, pine needles, blueberry, cowberry c. 20%, c. 45% roots and moss) 3% clastic debris sand grade up to c. 5cm. Base layer of charcoal from a forest fire.
5-10	Brownish grey 6/1	Sandy clay, with darker, pinkish mottling in profile. Coheres together into a ball, feels and sounds silty and deforms with cracks. Includes numerous clasts of sub-angular bed rock >0.5cm and <15cm. 90% clay/silt/sand 7% larger clasts 1 - 20cm 3% humic material, moss, roots.
10-20	Brown 4/4	Sandy soil, poorly sorted sub-angular igneous clasts (0.5cm - 10cm). 70% sand/clasts 30% organic material, dominantly tree roots.
20-30	Dull yellowish brown 5/4	Sandy clay, subrounded, poorly sorted clasts (0.5cm - 30cm) 78% Clay/silt/sand 20% Clasts 2% unidentifiable organic material.

TABLE 5.2: A detailed description of the soil horizons at the sampling location for Tree D

Depth (cm)	Colour	Description
0-20	Brownish black 2/1	Humic matter with plant material and occasional sub-rounded clasts of the bedrock. 98% Organic 65% Discernable plant material (pine needles, lichen, blueberry, cowberry c. 20%, c. 45% roots and moss) 2% clastic debris sand grade up to c. 5cm.
20-30	Brownish grey 5/1	Sandy clay, darker/pinkish mottling in profile. Coheres together into a ball, feels and sounds silty and deforms with cracks. Includes numerous clasts of poorly sorted, sub-angular igneous bedrock >0.5cm and <20cm. Also includes indistinct charcoal horizons. 80% clay/silt/sand 7% larger clasts 1 - 20cm 13% humic material, moss and roots.
30+	Dull brown 5/4	Sandy soil, poorly sorted sub-angular igneous clasts (0.5cm - 10cm). 98% sand/clasts 2% organic material, dominantly tree roots.

boundaries were identified. Soil samples of c.5g were taken from each horizon from three positions around the base of the remaining stump. Table 5.2 provides a detailed description of each of the identified soil horizons.

5.4 Laboratory preparation

5.4.1 Soil samples

The samples were transferred into paper trays and dried at 40°C for 1 week. 5g sub-samples from the same horizon for each of the three sample pits from each site were amalgamated to average out localised differences in soil chemistry. The rest of the samples were archived. The average samples for each horizon at each site were passed through a 2mm sieve. 10g of the >2mm fraction were finely ground in an agate ball grinder, and the remaining sample was archived. The samples were prepared for solutions analysis via a standard soil digestion procedure (see Appendix E).

5.4.2 Wood samples

On arrival in the lab, samples were removed from packaging and placed in an oven at 35°C for 24 hours to remove excess moisture. They were then left to dry in a cool, aerated room for two weeks. Once sufficiently dry to sub-sample, samples were chiselled in half and then sawed into half discs of c. 0.5cm. The discs were then placed upright in a rack to allow maximum air circulation and dried at room temperature for a further month. Whilst the ideal would have been to measure, dissect and analyse the samples straight from frozen (or better still whilst frozen), the lengthy preparation time required for solutions ICP-MS would have made this impractical. It was decided to aim to remove excess moisture from the samples as quickly as possible in order to minimise migration of sap and critically to prevent growth of mould. Some radial cracking occurred but this did not detract from the chemical work. There was no growth of mould.

Measurement - Five sub-samples from the 1.5m sample for each of the two trees were taken for measurement (see figures 5.2 and 5.3 for examples). This was so an average ring width pattern could be generated for each tree. The samples were sanded to highlight the ring pattern for ease and precision of measurement. Two radii were measured from opposing ends of each sub-sample. Ring width patterns were measured using a Lin-Tab measuring platform driven by TSAP (Time Series Analysis and Presentation) version 3.5 [1]. An average was taken of the ten radii and used to produce the ring width measurement curve for each tree. In the same way, average width measurements were generated for the early and late wood width of individual years for the same sequences. The aim was to compare the total yearly growth pattern with the sub-annual growth pattern and both of these with temperature, precipitation and chemical data in order to explore specific physiological relationships.

Dissection and preparation for analysis - Three sub-samples were taken from the base, the middle and the top of the 1.5m sample for each tree. The sample surfaces were cleaned by the removal of the uppermost layer of cells with a sterile steel razor blade. Samples were marked with an indentation made with a steel pin at ten year intervals, working backwards from the year of felling and last complete growth increment (2002). Under ×20 magnification individual tree rings were dissected from the main sample. The resulting triplicates of each ring were amalgamated in one labelled sample bag and mixed thoroughly; where necessary the rings were crushed using an acid washed agate pestle and mortar, into smaller fragments with the aim of standardising the size of all sample particles for each sequence. Sequences of individual tree rings were dissected from 2002 - 1800 for Tree C and 2002 - 1930 for Tree D. A sub-sample of the complete sequence of each was retained, cleaned and mounted ready for laser ablation. In order to test variation in a single tree ring, the whole circumference of one ring was dissected from a 1 cm cross section of Tree C taken at 1.5m from the base. This produced c. 8g of sample which was quartered into replicates. This was repeated for a ten year sequence from the sapwood (1986 - 1996) and a ten year sequence from the heartwood (1878 - 1888) so that further comparisons could be made between the chemistry of the two zones. In order to explore chemical differences in the same tree ring at different heights in the tree, a further two ten year sequences were dissected from the same ten sapwood and ten heartwood rings 2.5m and 3.5m from the base of Tree C.

Shavings from the cleaning process were weighed into 25ml Pyrex beakers, and ashed at a variety of temperatures in order to determine the best temperature / time combination for effective ashing. It was found that 400°C for 7 hours produced the best results. All samples were ashed, then $1\mu l$ of Aristar ultra pure concentrated HNO_3 and 5mls of UHQ were added. This produced almost total dissolution, however occasional samples contained fine undigested carbonised fibres. A filtration step was added to the procedure to counter this and all samples were passed through Whatman 45, ashless filter papers. Samples were prepared in batches of forty, and cold stored until analysis.

5.5 Environmental signatures to trace

Any potential environmental signatures to be found in the tree rings would have to be discernable against the tree's natural uptake patterns from pith to bark. Following an exploration into the physiological distribution of elements within the samples, and a consideration of site specific inputs such as soil chemistry, an attempt was made to ascertain if any type of environmental signature could be determined within the dated elemental sequences. Known dated sequences of potentially traceable environmental events or changes will be considered in this section. This will begin with localised events such as forest fires and fertilisation, and go on to consider temperature and precipitation, known patterns of anthropogenic pollution and local and global volcanism.

5.5.1 Forest fires and fertilisation

Information availability on factors specifically affecting the sampling site was important in the choice of sampling location. At Siljansfors Experimental Forest, dates of known forest fires are 1708, 1733, 1777, 1824 and 1857. Of these only 1857 and 1824 are covered by the sampled

---
[1] © Frank Rinn, Heidelberg - see http://www.frankrinn.com/Products/Tsap.htm.

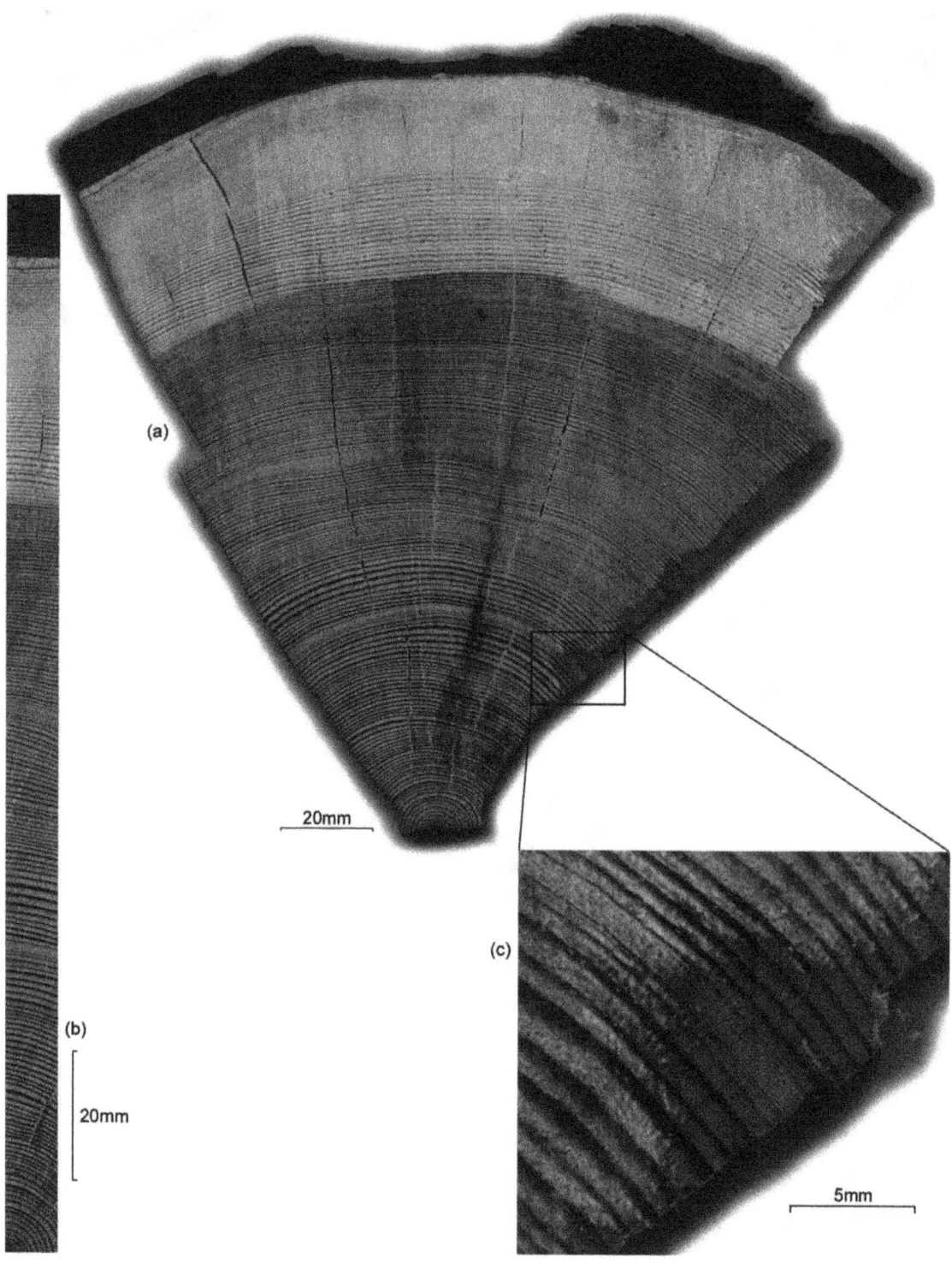

FIGURE 5.2: Tree C detail. A: Typical sub-sample prior to dissection. B: A close up section of Tree C tree ring sequence from pith to bark. C: A close up of the narrow ring event associated with the 1824 forest fire.

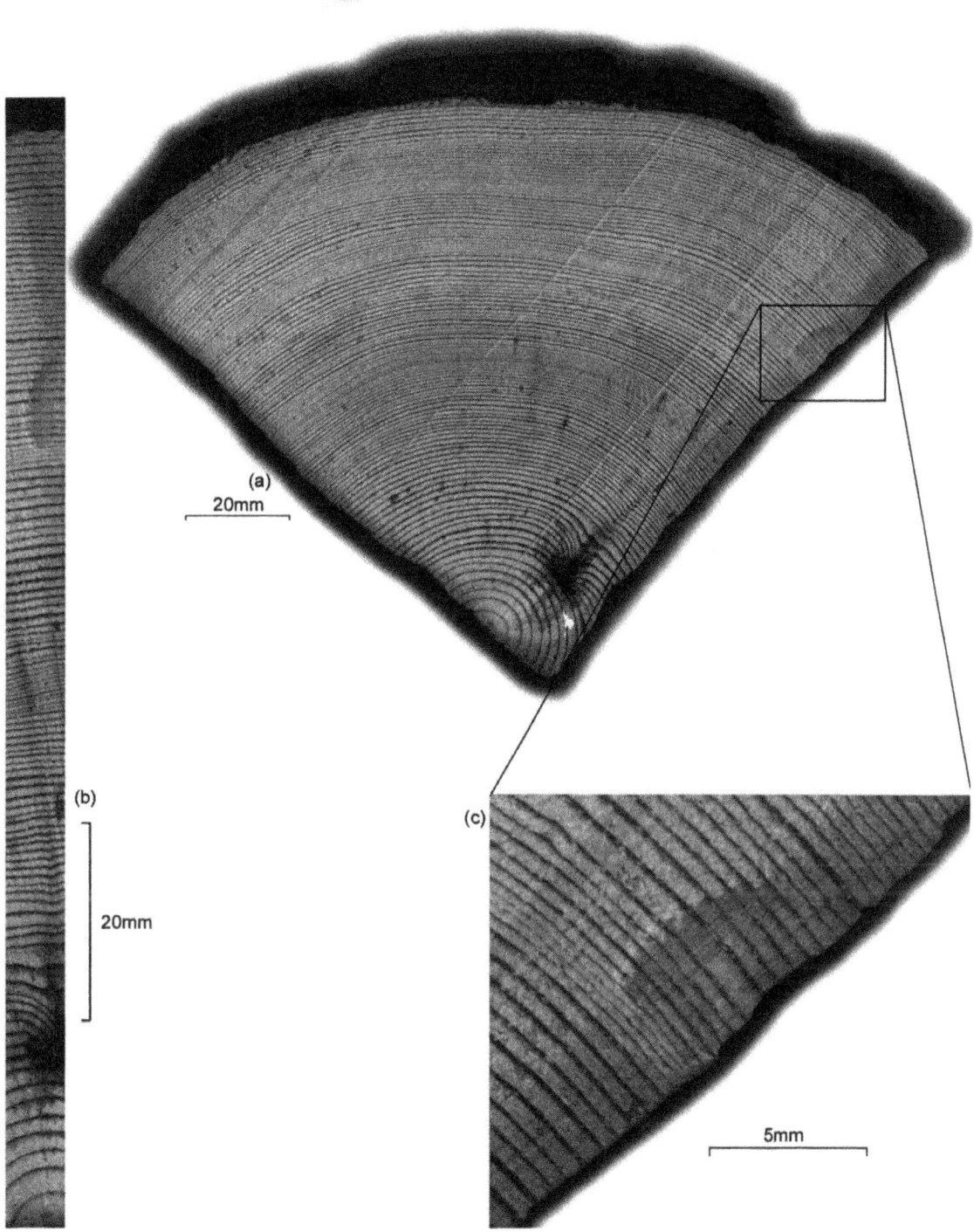

FIGURE 5.3: Tree D detail. A: Typical sub-sample prior to dissection. B: A close up section of Tree D tree ring sequence from pith to bark. C: A close up of the wider rings associated with fertilisation in 1967.

sequence of tree rings. In terms of fertilisation, only Tree D was fertilised in 1967 (see section 5.3.2). Both fires and fertilisation could reasonably be expected to have had some impact on the tree-ring sequence in terms of both growth and chemistry. In addition to the possibility that the growth rings may show suppressed growth in association with the forest fires and increased growth as a response to fertilisation, there may also be chemical changes associated with these events. For example, studies on the impact of similar fertilisation on soils in South West Sweden [106] have highlighted a resulting increase in extractable Ca and Mg lasting around four years after application of the fertiliser. It might therefore be anticipated that such an event would lead to an increase of these elements within the growth rings of a tree growing in such a soil. Similarly, Bondietti et al. [17] have found increases of Al, Ca and Mg in tree rings associated with the after effects of forest fires.

5.5.2 Temperature / precipitation

Detailed temperature and precipitation data specific to Siljansfors Experimental Forest were obtained from the Swedish University of Agricultural Sciences, spanning 1922-2002. Whilst these data only cover a short section of the total sequence of tree rings analysed, the eighty years provided were sufficient to gain an understanding of the degree of agreement with data sets generated for the corresponding annual growth increments. It was important to obtain data as local to the growth environment of the trees as possible in order to obtain the most likely possible match. According to Kindbom et al. [78], the amount of deposition of different compounds from the atmosphere is highly dependent on the amount of precipitation. Therefore one possible effect on tree ring chemistry could be the dilution or concentration of environmental deposition within annual rings in relation to the amount of precipitation. On the other hand if a tree's growth is enhanced by wetter weather it may be the case that elemental uptake is increased for years with less rain although that uptake may not necessarily relate to deposition but to root uptake from the soil. If this is the case it could then be argued that dryer years would be more representative of environmental deposition as this would be more concentrated and directly related to uptake from the air (an argument for selecting trees from very arid environments for further analysis). Temperature may accelerate up-take of pollutants with warmth speeding chemical reactions in the soil and on the bark surface.

5.5.3 Anthropogenic pollution

The study site was chosen because it was relatively unpolluted. However, possible pollution signals to detect could include general trends in background air pollution and smelting activity at Falun copper mine, 80 kilometres away.

Table 5.3 shows examples of the types of trends recorded over the last 100 years for various background environmental pollutants. Data have been compiled from a key Swedish pollution study [78] and several other sources [25, 42, 72, 131, 116, 120].

Falun copper mine is one of the oldest in Europe and produced two thirds of world copper at its peak in the 17th century. According to Ek et al. [40], emissions of sulphur dioxide from Falun since 1800 have seen an overall decrease. Within that trend, the twenty five year average is highest between 1800 -1825 and 1850 - 1875. It is slightly lower for 1825 - 1850 and 1875 - 1900, with much lower trends observed for 1900 - 1950, before rising slightly between 1950 - 1975 and tailing off dramatically from 1975 - 2000 following closure in 1993. This deposition is noted to have had quite marked effects on the immediately surrounding forests in terms of defoliation and decline. There is the possibility that the impact of these trends may therefore be detected in Siljanfors trees in terms of a rise or fall in particular elements ether due to fluctuating soil acidity or reduced functionality of the tree. In terms of heavy metal pollution, Ek et al. [40] note that soil pollution of the area immediately surrounding the mine has been fairly limited due to the local topography and the fact that emissions were largely at ground level. They suggest that the main polluted areas are up to twelve kilometres to the north west and south east of the mine. Given this, it is perhaps unlikely that any record of smelting operations will be found in the trees at Siljanfors, especially in terms of the most obvious by-product: Cu, which has been found to be dominantly emitted in larger

TABLE 5.3: Background trends in environmental pollutants (2001-1900). Data quoted for the whole of Sweden or where '†'; closest data collection point given to Siljanfors. Data from the Greenland ice cores is marked with '‡'

Year(s)	Pollution Record
1900-1910 ‡	SO_4 concentrations in Greenland ice cores begin to rise reflecting emissions from fossil fuel combustion
1930's ‡	A break in SO_4 deposition observed in Greenland ice cores
1940's-1960's ‡	A sharp increase in SO_4 deposition culminating in the 1960's
1949-1985	Soil pH increase accompanied by loss of exchangeable Na, K, Mg and Zn, due to reduced atmospheric acidic deposition.
1950-1989 ‡	Increase in NO_3 deposition
1968-2001	Heavy metals decrease gradually and evenly, reflecting a general improvement of air quality due to decreasing dust emission from industry and fossil fuel emissions. Pb showing the largest decrease followed by Fe, Cd, Ni, Hg, V, Cu and Zn. Zn has reached a natural baseline.
1970-1995 †	Heavy metals decrease, Pb showing the strongest decrease followed by Cr, Fe, Cd, Ni, Hg, Cu and Zn
1979-1998	Soot decrease
1980- ‡	A decreasing trend in sulphate deposition
1980-1993	Particulate S concentrations decrease 30 - 70%
1980-1993	SO_2-S decrease of between 40% and 80%
1980-1998	NH_3, S and NO_x emissions decrease
1981	Lower than average soot deposition
1982	Peak in SO_2-S concentrations
1982	Peak in soot deposition (up to $9\mu g/m^3$)
1983	Lower than average soot deposition
1983	Lower than average SO_2-S
1983	Lower than average particulate SO_4-S
1983-1998 †	Mg fairly constant as are Na and Cl
1985-1987	Steep SO_2-S decrease
1985-1998	Particulate S decrease from air
1986-1998 †	As, Cd, Ni and Pb vary from year to year, but decrease overall
1987 †twocolumn	Lower than average Mg, Na and Cl
1987 †	Fall in concentrations of NO_x- N and NH_4- N
1987-1998 †	NO_x- N concentrations decrease
1988-1998 †	Lower environmental acidity, (more H^+)
1989 †	Peak in concentrations of NO_x- N and NH_4- N
1990	Continuous steep SO_2-S decrease
1990-1996	Airborne Hg decrease
1991	Ozone unusually low
1998	Ozone unusually low
1992	Lower than average particulate SO_4-S
1995-1996 †	Higher than average Hg levels

particles which can not be transported far from the source [40]. It may however, be possible that some trace of more readily transportable emissions such as Zn, Cd and especially, Pb (which was found in its highest concentrations in a north south direction from the smelter) may be found in the tree rings. A further consideration as to whether this type of more long range impact can be expected is wind direction. According to Jönsson and Holmquist [75] the dominant wind direction in southern Sweden since around 1860 (when it switched from a bi-directional, west-east, continental flow pattern to a westerly oceanic one) has been westerly. This is generally the case at Siljansfors where the dominant direction is south or south west (Christer Karleson, pers. comm. 2003). The fact that Falun lays south east of Siljansfors further suggests that it is unlikely any history of pollution for the mine will be detected.

So to summarize, the types of elemental trend resulting from anthropogenic pollution which one might look for in the tree ring record from Siljanfors are:

From the early 1900's a rise in certain elements relating to increased industrial activity, with a break in this trend in 1930. In the 1940's an intensified increase, then a steady decrease from 1970, more marked from 1980. There may also be evidence of higher levels of elements such as Pb, Zn, and Cd in 1800-1825 and 1850-1875, with lower levels for 1900-1950 and the lowest levels from 1975 onwards. It might also be possible to detect the closure of Falun mine in 1993.

5.5.4 Volcanic eruptions

Against this background, there is the possibility of searching for an answer to the main research question, namely, can any chemical signature be found to link volcanic eruptions to tree ring growth years? It was decided to look for correlations between element patterns and eruptions on both a major global, and more minor continental scale.

Global scale - Table 5.4 shows the major (VEI 5+) volcanic eruptions of the last 200 years. Of these, Tambora (1815), Krakatoa (1883) and possibly Santa Maria (1902) leave acidity spikes in the Greenland ice cores [35]. More recently Mt Pinatubo (1991) has been added to this list.

Continental scale - Iceland provides a well documented source of more localised volcanism. Whilst most volcanism in Iceland is low level, effusive, and so most unlikely to be detected in Swedish tree rings, there have been occasions (e.g. the Laki fissure eruption [51]) where the impact of Icelandic volcanism has been felt across Europe. Also several instances of Icelandic volcanic ash being detected in lake sequences in central Sweden [175]. The studied sequences do not go back as far as Laki, however, they do cover many eruptive episodes (see table 5.13 on page 138)) and it is just possible that the relatively close proximity to the study site may make certain events detectable. Of the dates shown in table 5.13 a few can be highlighted as being the more probable to show some connection with tree ring chemistry (see table 5.5).

The fact that many overlaps exist between these eruption dates, dates of major volcanism and dates for pollution events highlight just how complex the potential signatures in tree rings could be, how hard they may be to tease out, and just how much conjecture is involved in any interpretation.

5.6 Results

Element concentration patterns were generated for forty seven different elements (Li, Be, Al, Ca, Ti, V, Cr, Mn, Fe, Co, Ni, Cu, Zn, As, Selenium (Se), Rb, Sr, Y, Zr, Rh, Ag, Cd, Sn, Sb, Cs, Ba, La, Ce, Pr, Nd, Sm, Eu, Gd, Tb, Ho, Er, Tm, Ytterbium (Yb), Lu, Hafnium (Hf), Pt, Au, Hg, Pb, Bi, Th, U), for one sequence of two hundred individual years (Tree C) and one sequence of seventy individual years (Tree D). In addition, a further hundred samples were prepared from the heartwood and sapwood in order to explore the chemistry of single tree rings at various heights in the trunk of Tree C. This produced a total of 17390 concentration values (564 graphed sequences) for analysis. This quantity of data can not be included in the text and only selected data sets will be used to represent key points. In addition to

TABLE 5.4: Major volcanic eruptions of the last 200 years. Compiled from Simkin and Siebert [142].

Year	Month	Eruption	Location	Explosivity
1991	June	Pinatubo	Luzon, Philippines	VEI 6
1991	August	Cerro Hudson	Chile	VEI 5
1980	May	St. Helens	Washington, US	VEI 5
1956	March	Bezymianny	Kamchatka	VEI 5
1932	April	Azul Cerro (Quizapu)	Chile	VEI 5+
1907	March	Ksudach	Kamchatka	VEI 5
1902	October	Santa Maria	Guatemala	VEI 6?
1883	August	Krakatau	Indonesia	VEI 6
1853	December	Chikurachki	Kurile Island	VEI 5?
1835	January	Cosiguina	Nicaragua	VEI 5
1822	October	Galunggung	Java	VEI 5
1815	April	Tambora	Lesser Sunda Island	VEI 7
1800	January	St. Helens	Washington	VEI 5

the concentrations data, complimentary data sets were also generated in relation to basic soil chemistry and the ring width patterns for the sequence. The results have been split into three sections, 'Endogenous controls', 'Local exogenous input' and 'General exogenous inputs'. The first section will present and discuss data relating to the internal physiology of the trees in an attempt to gain a better understanding of the context for further interpretations. The second section will relate to the observable influence of site specific local inputs to the tree chemical system. The third section will relate to the possibility for detecting the influence of wider ranging environmental signatures encapsulated in individual tree rings.

5.6.1 Endogenous controls

Any possibility of identifying a pollution or volcano induced signature relies upon first gaining as broad as possible an understanding of the control of the underlying physiology of the tree on the distribution of element concentrations within the xylem. It also depends on the potential reproducibility of the data sets for each element. This section considers elemental variation within single tree rings at various heights and positions in the xylem. In doing so it highlights the need for a careful approach to the interpretation of various elemental patterns and provides a guide to the reliability and reproducibility of certain patterns. In a similar way the degree of agreement between elemental patterns in the two trees is used to determine which elements are more suitable for the study of environmental signatures, and consideration is given to which of the more replicable patterns are due to physiological pith to bark orientation or changes at the boundary between sapwood and heartwood. Finally, as a measure of the reliability of the data sets as a whole (as well as the degree of replication to expect between trees) the average concentrations of each element for each tree are compared.

Variation in a single tree ring - Variation in a single tree ring at 1.5m from the base of Tree C was calculated. The percentage standard deviation for each element from four replicate samples of the same tree ring, at the same height, were produced separately for ten different years in the heartwood and ten different years in the sapwood. The ten values for each were then averaged and a percentage standard deviation was calculated to show the average degree of variation for each different element in a year. An example of the data produced is given in table 5.6, for both the heartwood and the sapwood. It shows that the degree of variation displayed by different elements in the same tree ring, varies greatly in both the heartwood and sapwood. Sr is the least variable element in both the heartwood and the sapwood, on average varying only 8% in a heartwood ring and 5% in a sapwood ring. V is highly variable (over 70%) in both heartwood and sap-

TABLE 5.5: Likeliest dates to show some trace of Icelandic volcanism in Swedish tree rings and the reasoning behind these conclusions

Year(s)	Possible Reasons for seeing Icelandic volcanism in the tree rings
1991	Two reasonably large eruptions of Hekla, but not during the growth season. Possible influx of deposition the following spring due to snow melt.
1980 - 1986	A lot of volcanic activity in this period, some during the growth season.
1970	2 months of eruptions at peak growth time.
1948	13 months of eruptions.
1947	One major explosive eruption right at the beginning of the growth season.
1938	2 small eruptions in one year, some during the growth season.
1934	2 small eruptions in one year, some during the growth season.
1921-1924	Lots of small scale activity.
1919	Might add to any impact felt from 1918.
1918	Major explosive eruption lasting 23 days, although late in growing season, this might show up in snow melt the following spring.
1913	Growth season eruption.
1875	A major plinian eruption with tephra fall observed throughout central Norway and Sweden [155] just before the onset of the growth season.
1873	Major explosive eruption lasting 7 months covering the main growth season (However, under a glacier).
1864	28 months of activity, a fissure eruption which melted through a glacier.
1860-1864	Lots of activity, of reasonable explosivity.
1845	Major explosive eruptions for 7 months, misses main growth but could show up in snow melt the following spring.
1823	Eruption of reasonable explosivity during the growth season.

wood rings. Aside from these two examples the average variability for particular elements appears to be quite different in the two parts of the tree. For example, Ag is 19% variable in the heartwood, but a much greater, 63% variable in the sapwood. In an attempt to explain these various apparent degrees of heterogeneity, these results were considered in respect of known factors such as anatomical element associations and known mobility (see 2.8 and section 2.6.3). However, no clear associations could be found.

Table 5.7 allows comparison of the average variability of all elements between individual annual rings. These values show that overall the variations in the chemistry of single heartwood and sapwood rings are reasonably consistent. There is little difference between the degree of homogeneity of each ring, it is the elements within the rings which vary to different degrees.

In order to further explore the data presented in table 5.6 the percentage standard deviation for each element in single rings from the heartwood and sapwood were ranked in relation to the ascending order of the percentage standard deviation values for the heartwood (see table 5.8). The table also includes the average concentration of each element in the heartwood and the sapwood to show any relationship between concentration and heterogeneity. Ranking the data clearly illustrates the difference between the variability of the same element in the heartwood and the sapwood. As the ranked numbers for heartwood and sapwood generally see the smaller / larger numbers ranked side by side, (with the exception of Ti), it seems there is a broad correlation between the elements. However, with the exception of Sr and V, no other elements showed exactly the same amount of heterogeneity in

TABLE 5.6: Average percentage standard deviation of individual elements in a single ring from both the heartwood and sapwood of *Pinus sylvestris*. For this table, data are shown to illustrate the selected isotope by which the various elements were measured. Further data will be presented as elemental values only. Where 'nd', the isotope was not reliably determined in this particular tree ring.

Isotope	Heartwood	Sapwood	Isotope	Heartwood	Sapwood
^{7}Li	59	64	^{103}Rh	58	nd
^{27}Al	17	14	^{107}Ag	19	63
^{43}Ca	16	20	^{111}Cd	11	13
^{47}Ti	63	12	^{121}Sb	nd	61
^{51}V	80	70	^{133}Cs	76	43
^{52}Cr	50	62	^{137}Ba	12	13
^{55}Mn	47	24	^{139}La	17	8
^{57}Fe	21	41	^{140}Ce	20	19
^{59}Co	17	20	^{141}Pr	17	15
^{60}Ni	58	58	^{146}Nd	15	21
^{63}Cu	24	17	^{147}Sm	15	nd
^{66}Zn	10	18	^{151}Eu	15	18
^{75}As	99	61	^{157}Gd	27	nd
^{85}Rb	22	15	^{159}Tb	32	nd
^{88}Sr	8	5	^{165}Ho	54	nd
^{89}Y	17	5	^{208}Pb	52	55
^{90}Zr	93	25	^{238}U	74	nd

TABLE 5.7: The average overall variability for all elements in a single ring (the average percentage standard deviation of four replicates of each element in a single year), for the given heartwood and sapwood sequences.

Heartwood	Value	Sapwood	Value
1888	44	1996	48
1887	48	1995	41
1886	45	1994	62
1885	45	1993	43
1884	42	1992	41
1883	42	1991	49
1882	44	1990	34
1881	38	1989	39
1880	56	1988	33
1879	54	1987	43
1878	44	1986	39

both the heartwood and the sapwood. Even so it was concluded that the elements ranked with the smallest numbers were likely to produce the more replicable data sets. For all the elements, the calculated percentage standard deviation values indicate the percentage of a given concentration which could potentially be seen as the noise created by the heterogeneity of that particular element. There were no readily apparent relationships between the concentration of the elements and their heterogeneity. The average concentrations for each element in the heartwood and sapwood are relatively similar, however there was a trend towards slightly higher concentrations in the heartwood, presumably because of the presence of lignifying substances.

The final step in the investigation of the chemistry of individual tree rings was to compare the average concentrations for each year in the two sequences with the concentrations of the same years at 2.5m and 3.5m from the base of the tree, thus testing how much variation exists up and down the trunk. The percentage standard deviation was calculated for three replicates of the ten year sequences, one at 1.5m, one at 2.5m and one at 3.5m. This was compared with those values produced by three runs at the same height. The values were not found to be significantly different, i.e. the degree of variation for each element was similar whether sampled from the same height or three different heights in the tree. As observed previously in this section for samples taken from 1.5m, differences were also found between concentrations of particular elements between heartwood and sapwood at 2.5m and 3.5m. As at 1.5m, Li, Ti, Cu, Rb, Sb, As and Cs were higher in the sapwood than the heartwood at 2.5m and 3.5m. However, Se, Ag, Ba, La and Pb were higher in the sapwood at 3.5m, and Be, Sr, Rh, Ag, Sb, Ba, La and Pb were higher at 2.5m. The overall percentage of elements higher in the heartwood at 2.5m was 53% and 32% were higher in the sapwood. At 3.5m the percentage of elements higher in the heartwood was 62% and 25% were higher in the sapwood. The rest of the elements were approximately equal in both at both levels. This is interesting as 70% of elements were higher in the heartwood at 1.5m. It emphasises the irregularity of element distribution within the trunk, however the proportions are still split in favor of higher overall heartwood concentrations.

Within all the variations some significant patterns can be determined. An increase in concentrations was noted with increasing height up the trunk for Be, Ca, Fe, Co, Rb, Sr, Y, Zr, Rh, Ag, Cd, Sn, La, Pr, Sm, Eu, Gd, Tb, Dy, Er, Yb and Pb in the heartwood and Li, Ca, Fe, Co, Cd, Pr, Nd, Tb, Dy and U in the sapwood. A decrease with height was noted for Li, Ti and Cd, in the heartwood and Se, Rh, Eu, Hf, Au, Hg, Bi and Th in the sapwood. Of these only Ca, Fe, Co, Cd, Pr, Tb, and Dy increase in both the sapwood and the heartwood with increasing height in the trunk. Other elements showed no pattern. All these data emphasise the complexity of the chemistry of a single tree ring, and the care with which any potentially true signatures must be interpreted and replicated.

Correlating element patterns between the two trees - Having established that certain elements are more replicable than others, and that the heterogeneity of the tree rings will impart a reasonable level of noise to any derived elemental sequence, a further check as to which pattern may be used for interpretation is to compare patterns for individual elements in each of the two trees. Where the patterns for any of the detected elements showed little agreement between the two trees, it was concluded that the data did not supply a sufficiently sound basis for further interpretation. For the long term potential to reliably identify a potential signal for volcanism or pollution in any tree ring sequence, it must be possible to replicate a particular anomaly in trees of the same species for the same site. Elements which displayed no correspondence between the two trees included V, Cr, Ni, As, Se, Sn, Sb, Ho, Yb, and U. These elements were found to be highly variable in a single tree ring (see table 5.8 for examples). It seems that the uptake and distribution of these elements must be controlled by very specific tree or microenvironmental factors. An alternative hypothesis is that the particular element is simply not available in sufficient quantities within the annual growth period to produce any replicable data sets. Whatever the causal explanation, it can be concluded that these elements are not suitable for tracing environmental signatures in *Pinus sylvestris*. In addition to this, data for Fe were also disregarded as discrepancies between the proportions of the detected isotopes indicated some sort of atomic interference in

TABLE 5.8: Ranked percentage standard deviations per element in the heartwood (HW) and sapwood sequences (SW). Where 'nd' the element was not determined.

Isotope	Concn (ppm) HW	Concn (ppm) SW	%StDev HW	%StDev SW	Rank HW	Rank SW
Sr	8	5	2.84	1.8	1	1
Cd	11	13	0.2	0.1	2	5
Ba	12	13	6.3	3.25	3	4
Ca	16	20	200	100	4	9
Al	17	14	5	5.2	5	6
La	17	8	nd	0.01	6	2
Ag	19	63	0.2	0.1	7	11
Fe	21	41	30	20	8	12
Rb	22	15	0.6	1.7	9	7
Cu	24	17	0.37	0.7	10	8
Gd	27	nd	0.001	nd	11	18
Mn	47	24	nd	30	12	10
Cr	50	62	nd	0.1	13	14
Pb	52	55	0.2	0.1	14	15
Ni	58	58	0.2	0.1	15	16
Rh	58	nd	nd	0.012	16	17
Li	59	64	0.1	0.1	17	13
Ti	63	12	0.1	0.4	18	3
V	80	70	0.01	nd	19	19

the ICP-MS system for parts of the sequences produced.

Of the rest of the elements Li, Al, Ca, Ti, Mn, Co, Cu, Zn, Rb, Sr, Y, Zr, Rh, Ag, Cd, Cs, Ba, La, Ce, Pr, Nd, Sm, Eu, Gd, Tb, Au, and Pb, all display some degree of visual correlation for all or part of the sequence[2]. Of these the least convincing are Ca, Mn, Zr, Ag and Cd. Further interpretations were made using these elements and taking into consideration the likely level of noise for each. The next step was to explore the nature of the observed elemental concentration patterns in an attempt to ascertain whether they were likely to be reflecting known physiological effects, rather than some exogenous input.

Physiological patterns of chemical change have previously been identified at the boundary between the inner, non-functional 'heartwood' and the outer, living 'sapwood', (see section 2.5.1). Sequences were studied for changes directly associated with the boundary (c.1935 for Tree C, c.1948 for Tree D) in both trees. The majority of the elements showed no obvious change in either sequence. The exceptions to this were Li and Ca, which show a rise to a peak in concentrations at the boundary itself followed by a decline, Ti, Y, Rb, and (to a lesser extent) Cu, which increased after the boundary, and Al and Zn which both displayed a decrease after the boundary. An example of the heartwood / sapwood boundary change for Rb in the two trees can be seen in figure 5.4. In addition to specific changes at the heartwood sapwood boundary, certain elements have been shown to broadly increase or decrease from the pith of the tree to the bark in accordance with the age of the tree (see section 2.5.2). Elemental sequences were studied for both trees. Where the overall pattern from pith towards bark produced for the two trees was similar, then a pith to bark trend was assumed. Where Tree D simply follows Tree C for the whole sequence an external influence is assumed. In cases where the pattern from pith to bark and the overall trend of the duplicate sequence was the same, elemental changes were considered at annual resolution. If rises and falls in concen-

---

[2]Where *part of the sequence*, this refers to correspondingly high or low concentrations for the same years or a similar overall trend interrupted by events or physiological boundaries specific to the individual tree.

tration could be discerned at several points in both sequences the influence was considered to be exogenous, if the trend was more general, influences were taken to be endogenous. The following pith to bark trends were observed in the two samples. Ti, Cu and Rb appear to increase from pith to bark, however the data for Rb (figure 5.4) match sufficiently well on a point to point basis to be interpreted as the result of some external factor, in addition to the influence of the change from heartwood to sapwood. Co, Zn, Sr, Y, Rh, Ba, La, Ce, Pr, Nd, Sm, Eu, Gd and Tb all display a dominantly decreasing trend. Of these Co, Zn, Sr, Y, Ba, La, Ce, Pr, Nd, Sm, Eu, Gd and Tb are judged to result from external factors. Therefore only Ti, Cu, and Rh appear to have a true pith to bark trend (see figure 5.5).

Concentrations - Having compared the overall patterns of the elemental sequences for each tree, the average concentrations were found to be very similar for the two tree ring sequences (see table 5.9). This is as would be expected for two trees of the same species grown on the same substrate at close proximity at the same site. The fact that they do, indicates that the data sets are reliable, and validates the sampling and analytical procedures. Discrepancies between the concentrations of particular elements in each tree can be explained by the averaging out of various peaks and troughs unique to the separate data sets and determined by the individual physiology of each tree and the impact of unique events such as fertilisation and forest fires. The slightly larger number of concentrations which are higher in Tree C may relate to slightly higher concentrations of various elements in the soil at the site. Although this relationship has not been proven in section 5.6.2, it is worth noting that of the sixteen elements detected in the two soils, twelve were higher at Site C (table 5.12).

5.6.2 Element sequences and local exogenous inputs

Soils - The soil at the two tree sampling sites is developed on a locally derived glacial deposit which overlays a Pre-Cambrian tuff. This igneous bed rock is highly siliceous, with small quantities of Al, K, Fe, Na, Ca, Mn and Ti. The overlying soil concentrations for the different horizons at sites C and D are shown in table 5.10.

As might be expected, the two sites have similar concentrations of various elements in the soil relative to one another. The differences between the different horizons in the two soils show the contribution of the organic surface component, and to an extent the presence of the second, leached horizon at both sites. The slightly higher levels of elements (for example as Al, Ti, Fe and As) present in the humic layer at site C might indicate that this site is exposed to slightly more deposition of airborne pollutants than site D. This may simply be a function of the immediate topography or exposure of the site, or it could be that the ground cover was dominantly composed of lichen species which are known to soak up pollutants [131]. Observations made during sampling confirm the presence of more lichen at site C than at site D. If this site was more exposed to pollutants it is likely that tree C would show the best potential signature of chemical change.

The chemical data supports the field interpretation that the soil is a podzol, with more Al, Fe and Ni and less Ca observed in the leached horizon. The fact that chemical differences between the horizons are poorly defined and that the overall concentrations are low, confirms the correct selection of a mature, nutrient poor soil.

Average concentrations were calculated for various elements in the soil at each site. These are presented in table 5.11, along with the average elemental content of the corresponding wood sample in order to ascertain if any relationship could be observed. As no obvious relationship could be observed this was thought to reflect the fact that trees selectively extract the nutrients they require in the quantities they need from the soil. It was observed however, that the average concentrations for the various elements were very similar for the two trees and, as observed previously, the two soils.

Potential relationships between soil and tree elemental composition are further explored in table 5.12. Here Al, V, Cr, Fe, Co, Zn and Ba are shown to be high in both soil and Tree C, Ca, Ti and Mn were highest in soil C and Tree D and Sr and Ag were highest in soil C and the same in Trees C and D. All these associations may represent some sort of relationship between the soil and the wood content.

CHAPTER 5: CASE STUDY 1 - SWEDEN

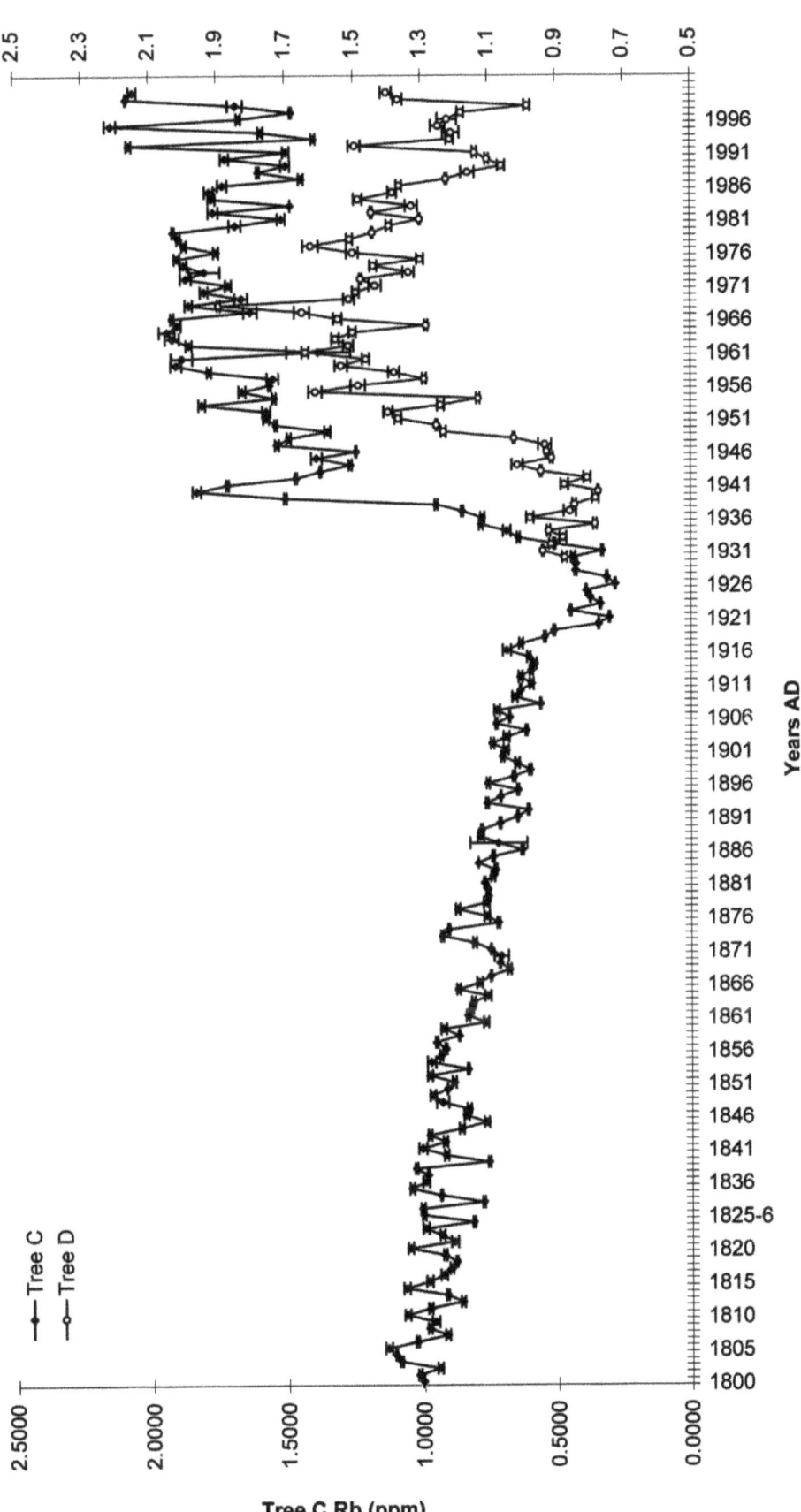

FIGURE 5.4: Rb in Tree C and Tree D, shows an increase from the heartwood / sapwood boundary c.1935 in Tree C, c. 1948 in Tree D

Chapter 5: Case Study 1 - Sweden

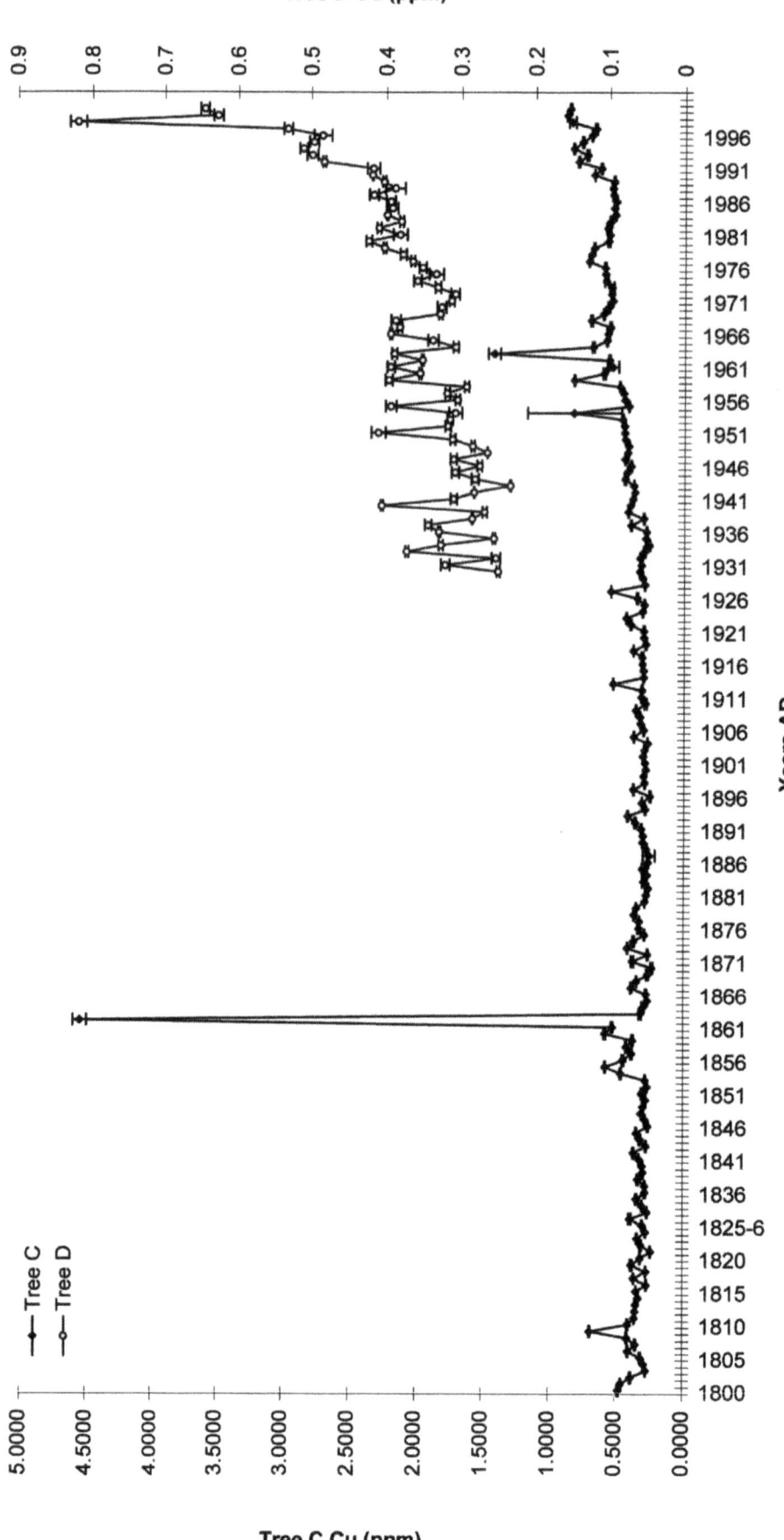

Figure 5.5: Cu in Tree C and Tree D, shows a gradual increase from pith to bark

TABLE 5.9: Average concentrations (ppm) per sequence for each element in trees C and D

Element	Mean Conc (ppm) Tree C	Tree D	Element	Mean Conc (ppm) Tree C	Tree D
Li	0.03	0.05	Ag	0.2	0.2
Al	4.5	3.9	Cd	0.17	0.14
Ca	163	189	Sn	0.003	0.002
Ti	0.13	0.15	Sb	0.004	0.001
V	0.017	0.003	Cs	0.004	0.003
Cr	0.02	0.01	Ba	4.8	4.6
Mn	19.9	30.3	La	0.027	0.045
Fe	31.2	23.1	Ce	0.03	0.04
Co	0.013	0.011	Pr	0.003	0.004
Ni	0.1	0.1	Nd	0.01	0.01
Cu	0.43	0.38	Sm	0.001	0.001
Zn	9.3	8.2	Eu	0.002	0.002
As	0.005	0.003	Gd	0.001	0.001
Se	0.006	0.005	Tb	0.0001	0.0001
Rb	1.06	1.31	Dy	0.0002	0.0002
Sr	2	2	Au	0.001	0.0007
Y	0.003	0.003	Hg	0.003	0.001
Zr	0.005	0.003	Pb	0.22	0.13
Rh	0.0005	0.0003			

TABLE 5.10: The concentrations (ppm) of the different soil horizons identified at sites C and D. Where 'nd', not determined.

Element	C0-5	C5-10	C10-20	C20-30	D0-10	D10-15	D15-30
Na	650	3600	5200	8300	460	4800	5700
Mg	540	1200	680	1700	340	470	1200
Al	6500	20900	14400	19600	2100	8900	41500
S	840	120	200	200	920	62	230
K	3980	19900	22500	27300	1460	21400	20300
Ca	1400	590	900	1300	1300	390	940
Ti	250	190	1700	1800	120	2000	1000
V	5.6	13	19	24	3.5	13	14
Cr	49	37	39	40	34	38	34
Mn	115	120	118	173	171	106	110
Fe	2800	6300	9600	10000	1200	5700	8500
Co	1.4	nd	nd	1.2	nd	0.59	0.43
Ni	5.4	3	4.2	5.7	4.2	4.6	5.4
Cu	9.3	3.5	5	2.9	6.7	5.8	5
Zn	59.7	44.6	141	33.4	54.9	29.1	31.4
As	1.7	nd	0.4	1.7	0.6	1.5	3.3
Sr	17.4	26.7	22.2	53.9	11.5	22.8	41.1
Ag	0.08	0.50	0.4	0.50	nd	0.50	0.05
Ba	62	140	170	290	35	150	210
Pb	34.6	23.7	34.6	20	35.5	20.1	29.9

Another way to identify potential relationships more clearly was through the calculation of ratios between each element in the two soils and each element in the two wood samples. The closer the ratio for the soils and the woods, the greater the likelihood that a relationship exists. Fe and Ni were shown to share the same ratio, suggesting that for these two elements a direct relationship does exist between the amount available in the soil and uptake into the wood. The final step taken to explore the data was the calculation of an 'uptake' factor from soil to wood. This was done by dividing the concentration of each element in the tree by the concentration in the soil. Hypothetically speaking the uptake by two trees of the same species from very similar sites could be expected to be approximately the same, and this was generally found to be true. The two trees were shown to have taken up the same degree of Al, Ti, Cr, Fe, Ni, and very similar levels of V, Cu, As and Pb. The majority of the other elements had been taken up more by Tree D, this may be explained by the fact that the soil at site D was fertilized in the 1960's. This could have had a buffering effect on the soil chemistry causing certain elements to occur in different speciation, and so becoming more available for the tree to take up.

5.6.3 Ring width measurements

The ring width measurement patterns for trees C and D are shown in figure 5.6 and figure 5.7. The total annual growth pattern is shown with measured divisions of the early and late wood. The purpose of presenting the measurement data in this way is to allow for comparisons to be made between the two growth divisions. This shows that for the majority of years good or bad growth is reflected by both increments and seasonal differences are not generally reflected in one increment or another. The pattern for Tree C is punctuated by several dramatic growth cessations. These are likely to be the result of specific events such as forest fires (see section 5.6.3). The first is short lived, from 1773, at its worst in 1774 and recovering after 1775. The second is the most marked, it commences in 1824 and recovery does not begin until around 1833, returning to normal in 1835. The third occurs around 1873, is at its worst in 1874 and slowly recovers to return to normal in 1881. Interspersing these very obvious events are less marked down turns in growth, such as 1850 to 1854, 1925[3] to 1935, and 1961 to 1965. There are also peaks in the growth pattern, again less marked than the major depressions. These occur around 1777, 1781-1783, 1790, 1799, 1806, 1816 to 1822, 1834, 1831 to 1832, 1846, 1860, 1882 to 1889, 1894-1896, 1915, 1940 and 1943. From around 1960 the pattern looses amplitude. The pattern for Tree D declines from 1880 to around 1912 but shows none of the really severe growth depletions observed in Tree C. It has peaks in 1915, 1924, 1940, 1943, 1949, 1953 to 1958, 1986 and 1995 to 1997. The most marked of these being that around 1968.

Figures 5.8 and 5.9 show how the section of the sequence covered by both trees compares for the two samples, and also to locally derived average yearly temperature and precipitation sequences. Figures 5.10 and 5.11 show the same measurements against a slightly refined data set for both temperature and precipitation. Here the averaged environmental data refers to the May to September growth season only, to test if the tree ring pattern shows less or more correlation with conditions at the specific time of growth as oppose to the yearly average. The correlation between the two tree ring width sequences is good. As would be expected, the two trees growing in close proximity to one another display a similar growth pattern - this demonstrates one of the basic principles of dendrochronology. Whilst overall the pattern matches throughout the sequence (with the exception of around 1968 where the results of fertilisation in 1967 are clearly seen in the sequence for Tree D), there is a notable difference in the width of the tree rings for the early part of the sequence. This suggests that Tree D had a slightly better start in life as a young tree and gradually accrued less annual growth tissue as it matured. The wider rings at the start of this sequence stand out as they are set against a mature section of the longer lived Tree C. If figure 5.6 is considered, it is however clear that Tree C did not benefit from such a favorable a start to life, as there is no marked widening of the rings towards the pith. One possible hypothesis to explain this is that according to Tree C's ring width pattern, the onset of growth for Tree D (in around 1880) was preceded by several very bad years starting from c. 1872. If

---
[3] worst in 1928

TABLE 5.11: The average concentrations (ppm) of the soils and within the trees from sites C and D. Concentrations are given in descending order for each element. Only elements detected for both soils and wood are included

Element	Soil C	Element	Soil D	Element	Tree C	Element	Tree D
Al	15300	Al	14400	Ca	160	Ca	190
Fe	7200	Fe	5100	Fe	31	Mn	30
Ti	1400	Ti	1100	Mn	20	Fe	23
Ca	1100	Ca	890	Zn	9.3	Zn	8.2
Ba	160	Ba	130	Ba	4.8	Ba	4.6
Mn	130	Mn	130	Al	4.5	Al	3.9
Zn	70	Zn	38	Sr	2	Sr	2
Cr	41	Cr	35	Cu	0.43	Cu	0.38
Sr	30	Pb	29	Pb	0.2	Ag	0.2
Pb	28	Sr	25	Ag	0.2	Ti	0.2
V	15	V	10	Ti	0.2	Pb	0.2
Cu	5.9	Cu	5.8	Ni	0.1	Ni	0.1
Ni	4.6	Ni	4.7	Cr	0.02	Co	0.011
Co	1.3	As	1.8	V	0.017	Cr	0.01
As	1.2	Co	0.5	Co	0.013	V	0.003
Ag	0.37	Ag	0.29	As	0.005	As	0.003

TABLE 5.12: A summary of relationships between soil chemistry and wood chemistry at sites C and D.

Element	Highest in soil	Highest in Wood	Ratio Soils	Ratio Wood	C Uptake	D Uptake
Al	C	C	1.1	1.2	0.001	0.001
Ca	C	D	1.2	0.9	0.155	0.213
Ti	C	D	1.3	0.9	0.001	0.001
V	C	C	1.5	5.7	0.002	0.001
Cr	C	C	1.2	2.0	0.001	0.001
Mn	C	D	1.0	0.7	0.152	0.237
Fe	C	C	1.4	1.4	0.005	0.005
Co	C	C	2.5	1.2	0.011	0.022
Ni	D	same	1.0	1.0	0.022	0.022
Cu	D	C	0.9	1.1	0.084	0.066
Zn	C	C	1.8	1.1	0.134	0.214
As	D	C	0.7	1.7	0.005	0.002
Sr	C	same	1.2	1.0	0.067	0.08
Ag	C	same	1.3	1.0	0.541	0.69
Ba	C	C	1.3	1.0	0.03	0.036
Pb	D	C	1.0	1.7	0.008	0.005

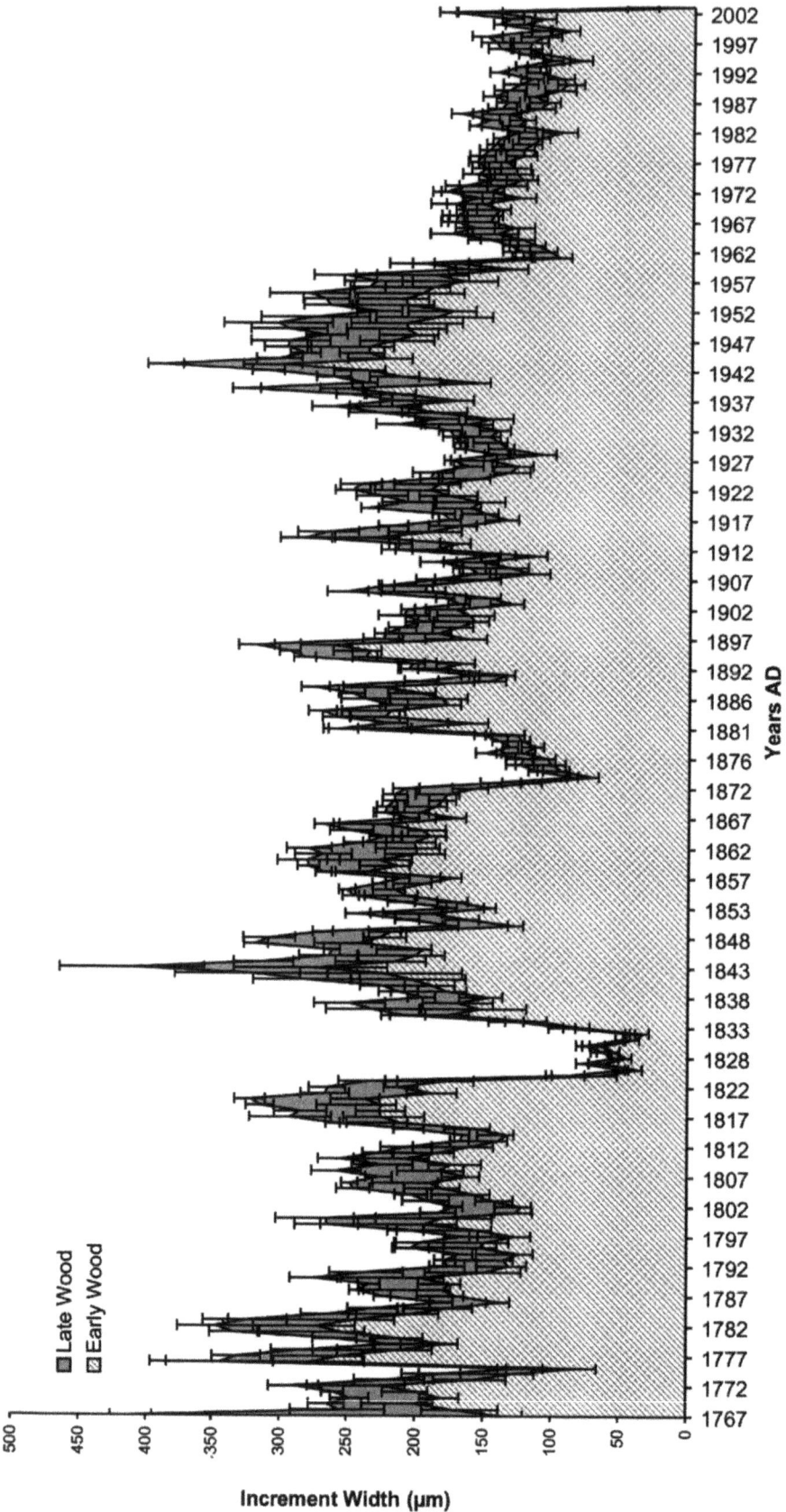

FIGURE 5.6: Early and late wood averages making up the average annual ring width pattern for Tree C

CHAPTER 5: CASE STUDY 1 - SWEDEN

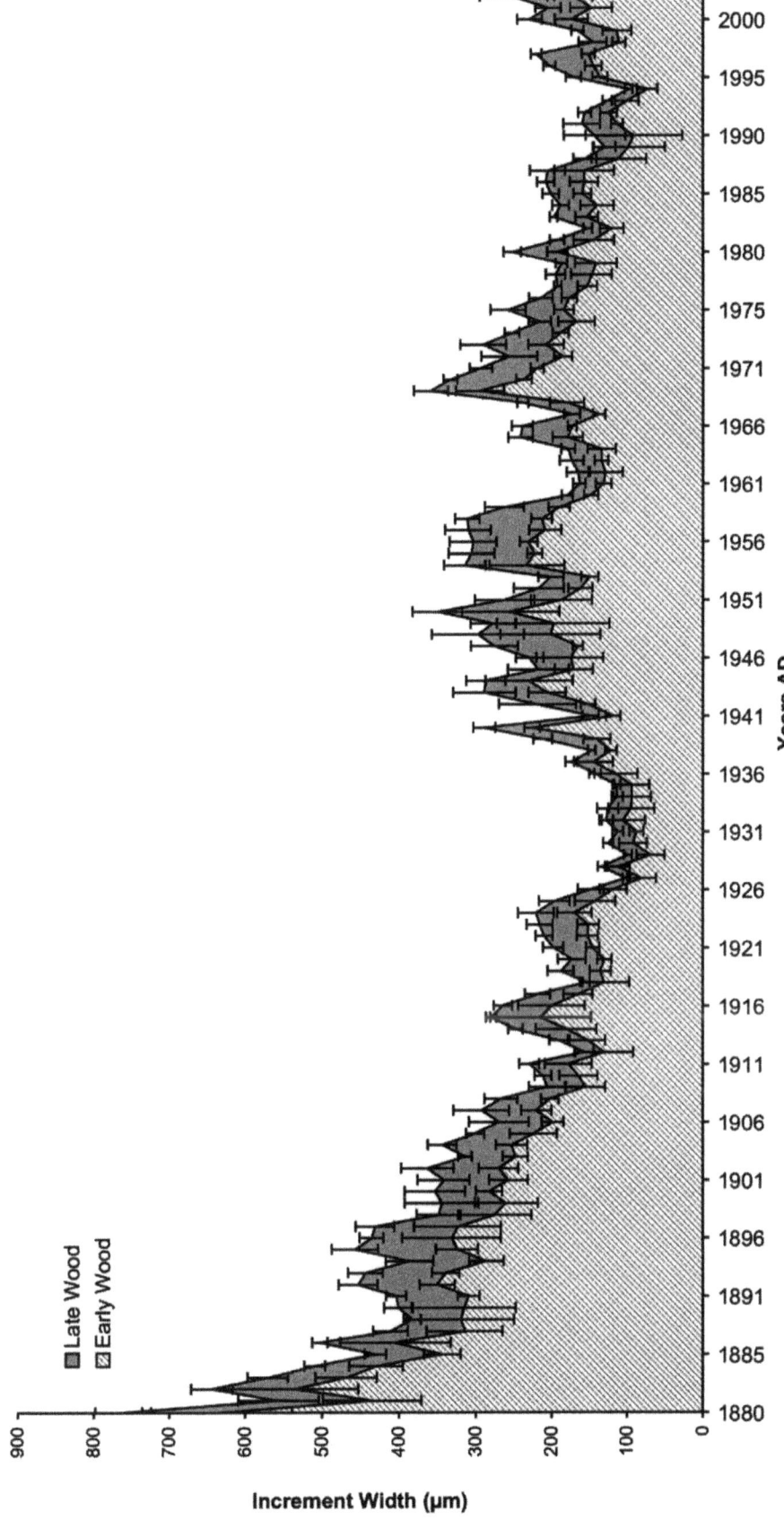

FIGURE 5.7: Early and late wood averages making up the average annual ring width pattern for Tree D

the cause of this narrow ring anomaly in Tree C was something such as an unrecorded forest fire, or particularly bad spate of climatic conditions then other trees may have died as a result leaving a gap in the canopy (and in the case of a fire, a layer of very fertile soil) and thus providing ideal growth conditions for Tree D.

The similar ring width pattern suggests that the two trees were responding to the same wide ranging environmental conditions and it was expected that correlations between ring width pattern and highs and lows of temperature and precipitation would be obvious. However this relationship was not found to be clearly demonstrated between the given data sets, even at growth season resolution. Figures 5.8, 5.9, 5.10 and 5.11 show that average yearly temperature / precipitation appears to have more baring on the ring width patterns than the previously identified May to September growth season. One possible reason for this is that the true length of the growth season may vary due to less or more favorable climatic conditions over the course of each year. This would produce a pattern which matches in both trees but does not directly correspond with known temperature or precipitation data. Apparent correlations with yearly average temperature (see figure 5.8) were identified in the early 1940's where lower temperatures correspond with narrow ring growth. This also occurs in 1952, however it should be noted that in the mid 1980's the lower temperatures recorded have no correlation with narrowing ring width. In terms of precipitation, the correlation is slightly more obvious with the growth season average (see figure 5.10). Here the lowest level of precipitation in the sequence (1959) corresponds with a narrowing of the tree rings, however, again there are other points in the sequence where this association is not made. Another possibility for the lack of obvious association between these data sets is that the trees are not growing in a marginal environment and annual average precipitation and temperature are relatively suitable for growth at all times, so that only the more extreme events really show up. Again the complexity of the tree ring growth environment is highlighted. In addition to this part of the analysis, element sequences for each of the two trees were plotted against the ring width, temperature and precipitation data sets (yearly and growth season averages). However, aside from the growth anomalies discussed in section 5.6.3 no clear correlations were evident in any of the data sets.

Sequences and forest fires / fertilisation - It should be noted that the 1824 fire had such an impact on the ring width pattern of tree C that it was not possible to obtain individual samples for the years 1827-1832 and 1825 and 1826. These years were therefore bulked together and analysed as two samples which may go some way to explaining the abrupt nature of the anomaly. The 1824 forest fire stands out as the most prominent chemical anomaly in any of the sequences. It is characterised by a rise in concentrations from 1824 up to 1832, then a fall in concentrations for 1833 and 1834, after which levels return largely to the same concentrations as before the event. This pattern is common to all the following elements; Al, Ca, Cr, Mn, Co, Ni, Zn, Sr, Y, Ag, Cd, Ba, La, Ce, Pr, Nd, Eu and Au. An example for Ba and Sr can be seen on figure 5.12, other examples can be seen on figure's 5.13, and 5.14.

At first examination, the 1857 forest fire is not recorded in the ring width or chemical pattern of the tree rings. However, the given date may be incorrect. Both the ring width and chemical patterns produce a similar (although less well defined) anomaly to that around 1824, from about 1849 to 1857. The given dates for the forest fires were derived by a forester named John R. Karlsson in 1921, who is thought to have counted the tree rings for each sequence without the aid of a microscope or measurement platform. There is therefore some possibility that he may have been out by a few years, especially if one considers the extremely narrow rings caused by the 1824 fire (see figure 5.2b), and the fact that for this study, eight years in this section were found too narrow to allow for separation, and indeed cannot readily be determined with the naked eye. If Karlsson were out by eight years, this would make his 1857 equal to 1851 according to the new measurements. However, consideration should also be given to the the major narrow ring anomaly at 1872/3 which does not recover until around 1881. Whatever event caused this impact on the life of the tree, the scale was very similar to the 1824 event. It seems odd that no forest fire is recorded in association with this event, where a series of small peaks occurs for Ba and several other elements. It is possible that this event was caused by a climatic event or even an

CHAPTER 5: CASE STUDY 1 - SWEDEN

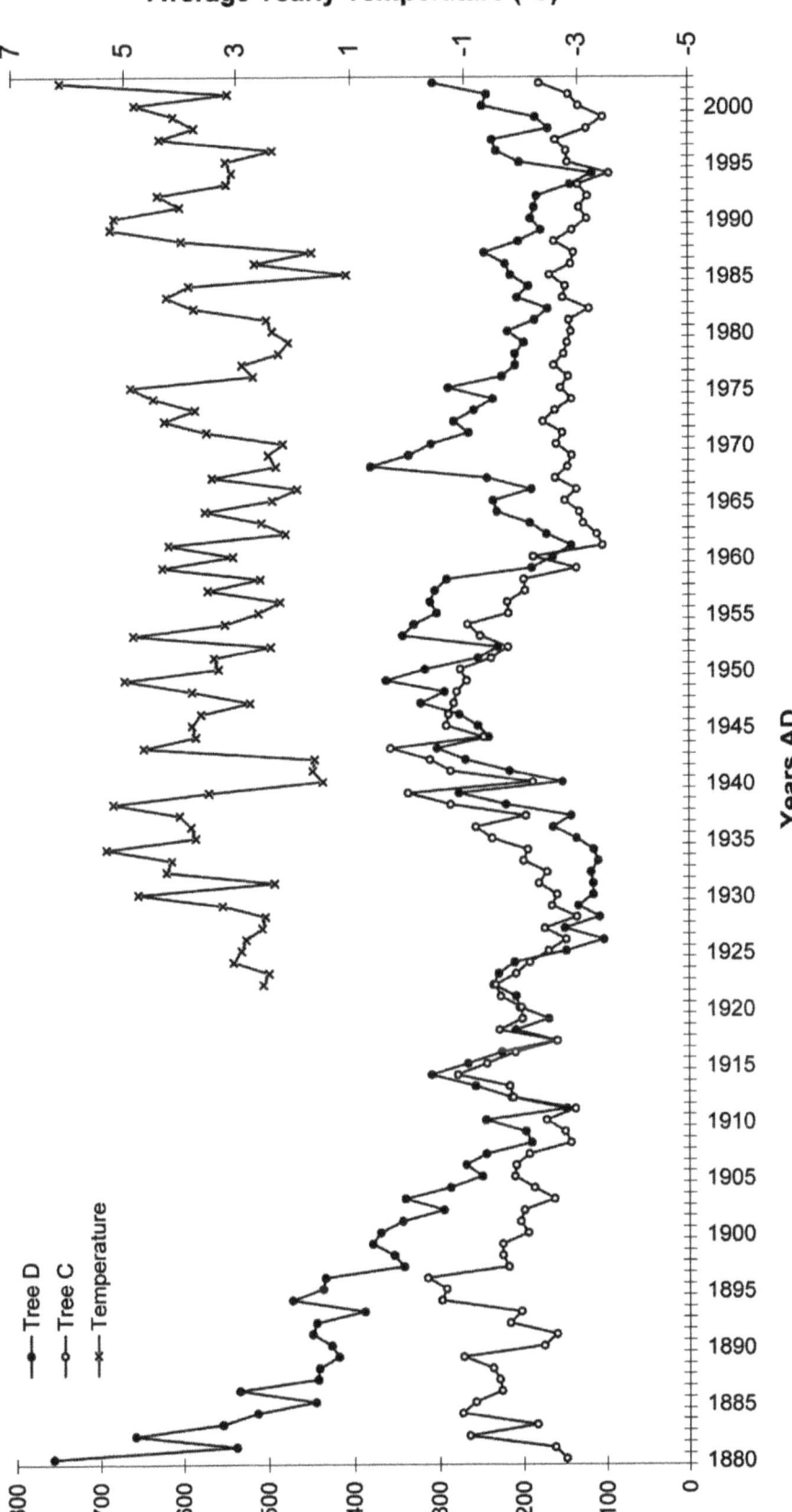

FIGURE 5.8: Sequence covered by Tree C and Tree D (1880 to 2002) against annual average temperature data (1922 to 2002)

CHAPTER 5: CASE STUDY 1 - SWEDEN

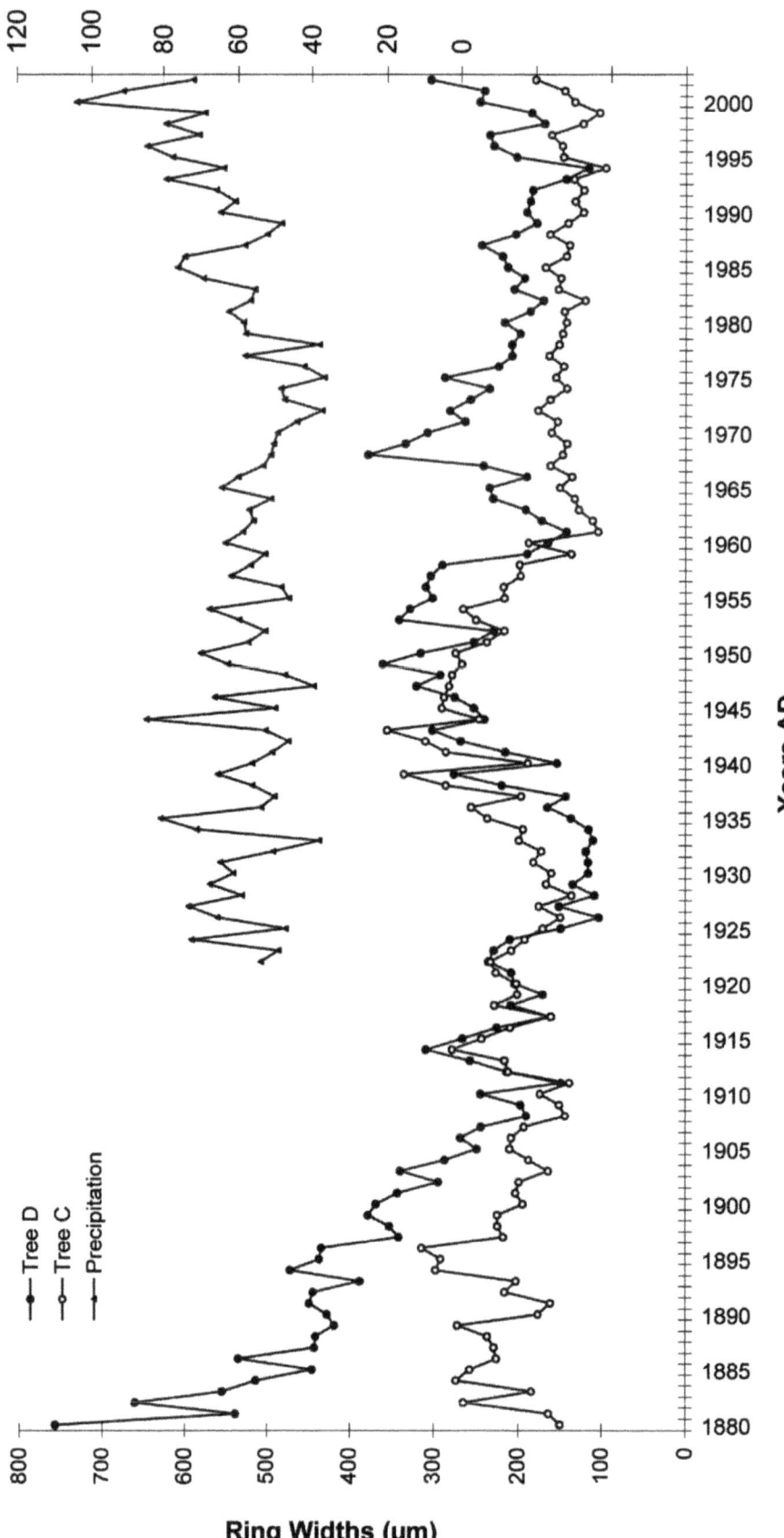

FIGURE 5.9: Sequence covered by Tree C and Tree D (1880 to 2002) against annual average precipitation data (1922 to 2002)

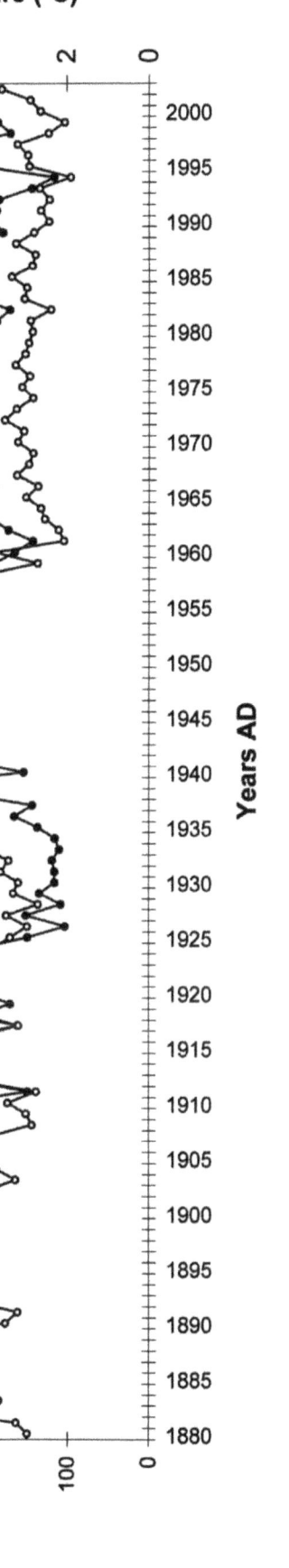

FIGURE 5.10: Sequence covered by Tree C and Tree D (1880 to 2002) against the average temperature data for the five month growth season (1922 to 2002)

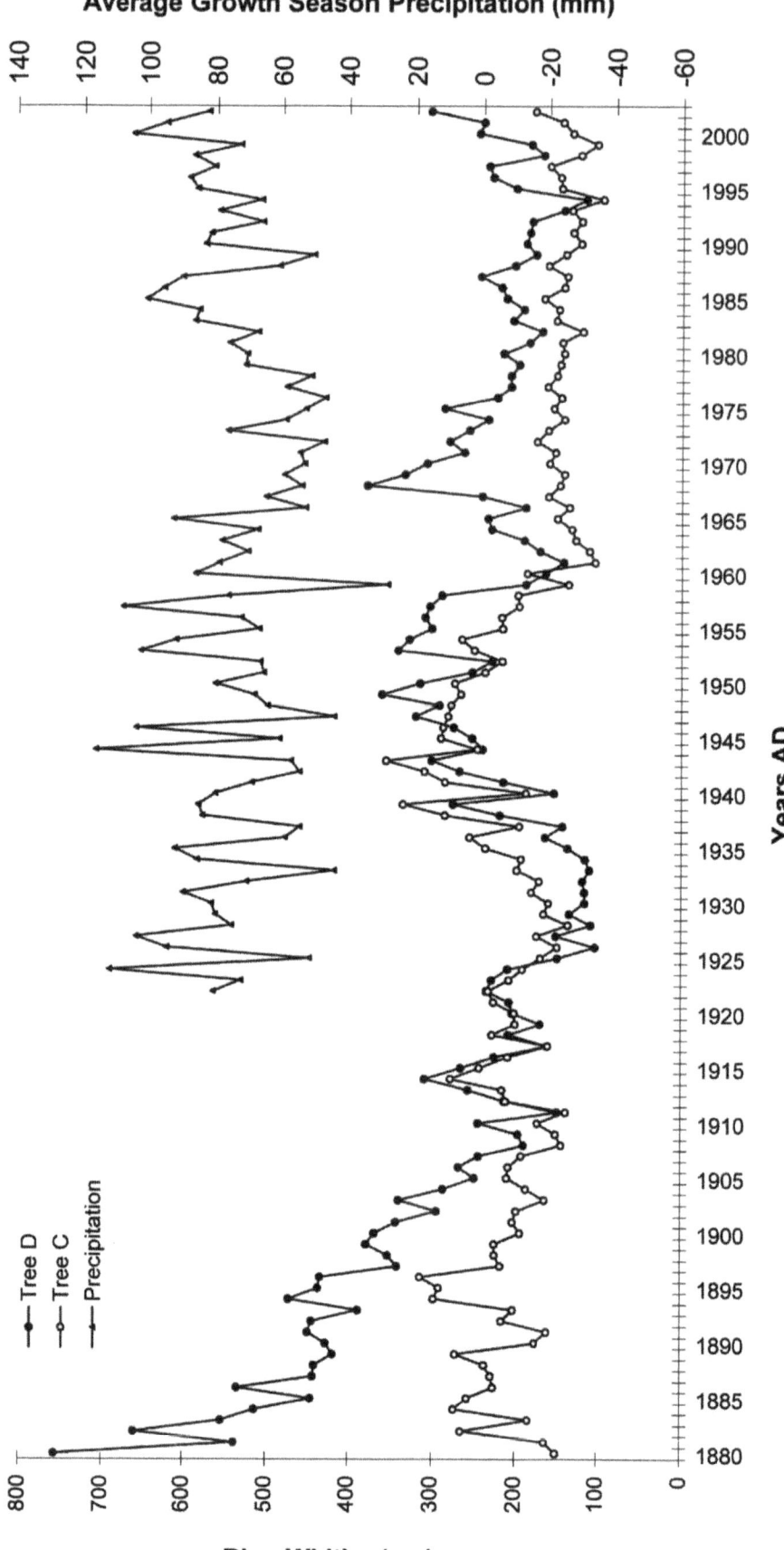

FIGURE 5.11: Sequence covered by Tree C and Tree D (1880 to 2002) against the average precipitation data for the five month growth season (1922 to 2002)

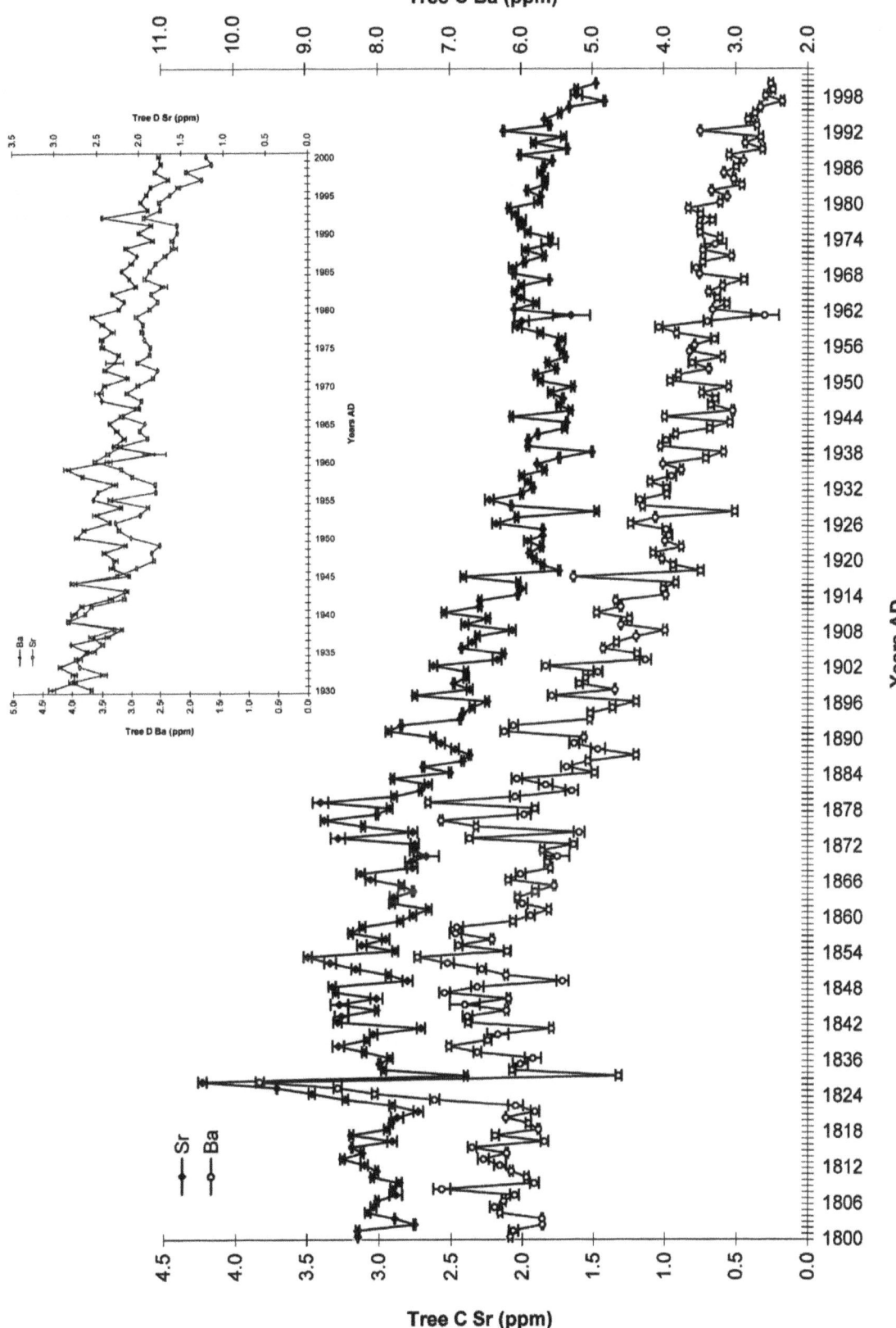

FIGURE 5.12: Sr and Ba in Tree C showing a marked anomaly around the 1824 forest fire narrow ring event. Agreement between these elements in Tree D is shown in the inset

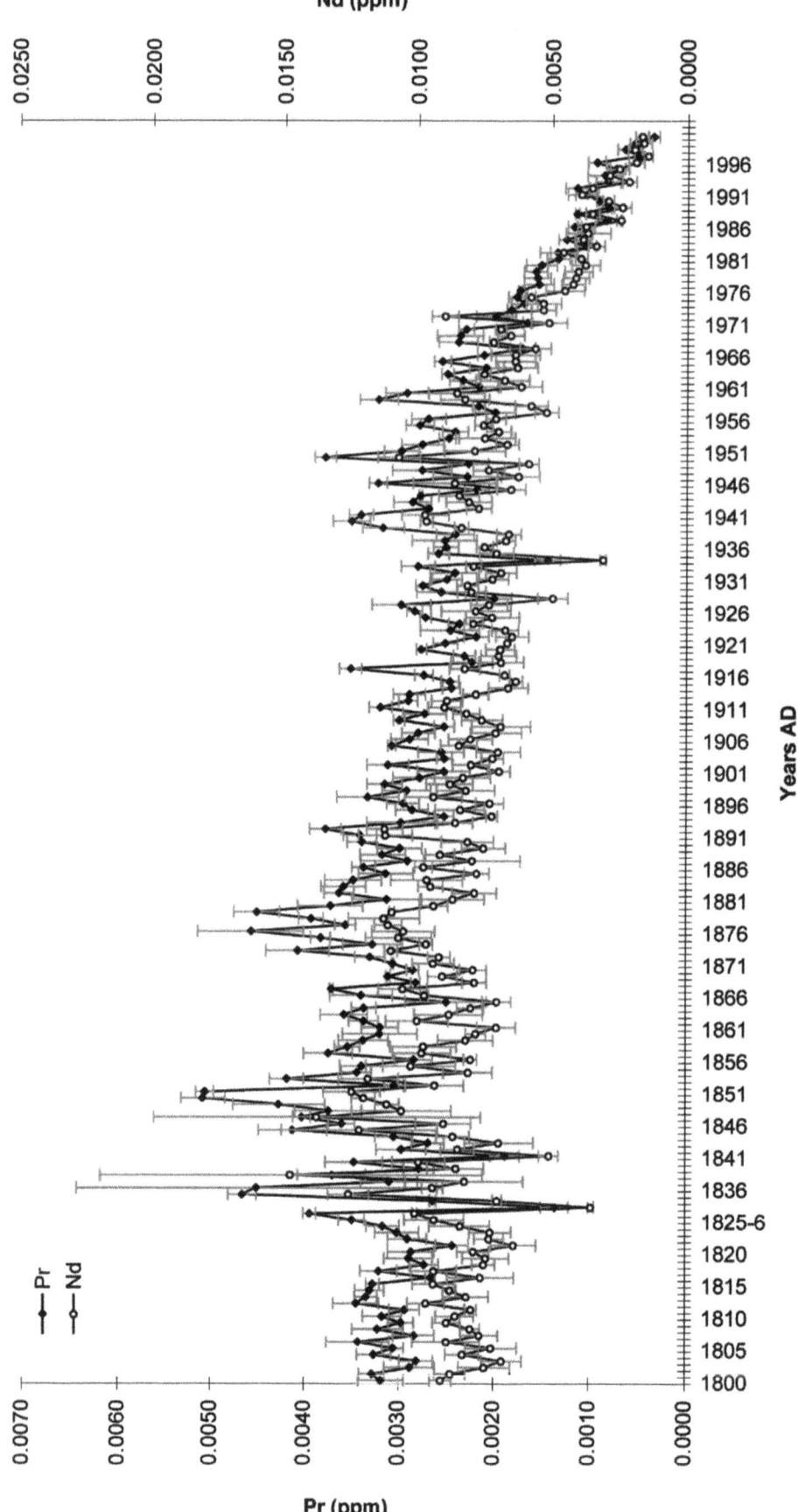

FIGURE 5.13: Nd and Pr in Tree C (1800 to 2002)

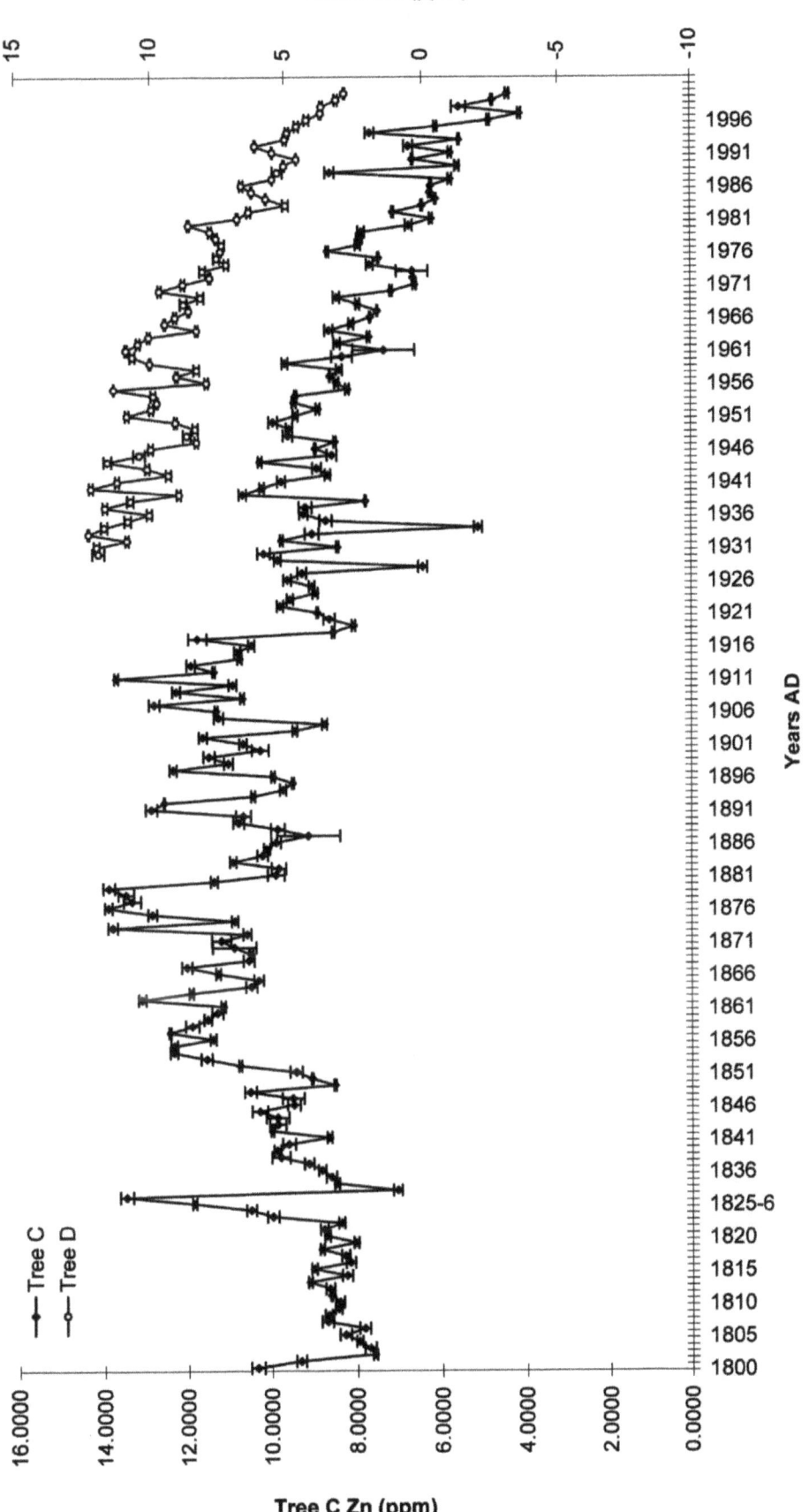

FIGURE 5.14: Zn in Trees C (1800 to 2002) and D (1930 to 2002)

eruption (see page 135 for further discussion), but it must be acknowledged that it is a far more likely candidate to be linked with a forest fire. In terms of the impact of fertilisation on Tree D, the main effect appears to be on ring width (see figure 5.3b and 5.7), with surprisingly little effect registering in the chemistry of the tree rings. The one element to show any real response is Cs which shows a definite increase following fertilisation (figure 5.15).

5.6.4 Element sequences and general exogenous inputs

Sequences and anthropogenic pollution - Graphs for each element for the two sequences were studied against the possible patterns to be identified for pollution summarised in Table 5.3. Whilst there are no obvious correlations with regard to the noted increases and decreases of certain pollutants, some parts of the more replicable element sequences were found to display similar trends to those noted in the table. The most convincing of these was a marked decline for many elements (including Sr, Ag, La, Ce, Pr, Nd, Gb, Tb) from around the early 1970's. An example of this is shown for Nd and Pr in Tree C in figure 5.13, and for La and Ce in Tree D in figure 5.16. In Tree D a rise occurs from 1991, this also occurs to less of an extent in Tree C (see inset figure 5.16). A possible explanation for this could be a sudden influx of these rare earth elements into the atmosphere.

Of the main elements noted in table 5.3, Na and K were not detected, Fe displayed instrumental interferences, Hg and V could not be effectively replicated in a single tree ring and Mg and Ni showed no correspondence between trees C and D in terms of the known pollution history. Pb, (figure 5.17) rises to a peak in the middle of the sequence from Tree C, possibly in association with a peak in turn of the century emissions, then declines from the late 1930's to the present day, broadly in accordance with the pollution history. The highest concentrations may reflect peaks in production at Falun, or possibly changes in wind direction bringing in vehicle pollution from another source.

The sequence from Tree D shows very similar concentrations and overall trend, with some individual peaks and troughs matching Tree C closely e.g. 1966 to 1971. This indicates that a true environmental trend is being detected. Zn and Cd display a gradual rise from the early 1800's and a gradual decline from the early 1900's for both sequences of samples. An example of this for Zn is shown in figure 5.14. It does not appear to relate to any particular pollution events except for the noted decline throughout the 1900's, steepening from around 1970. As with Pb, the two trees display similar concentrations and the overall trend of the pattern matches well. The decrease in concentrations for Tree C around 1935, and Tree D around 1948, is thought to have a physiological explanation (see section 2.5.1), where as the anomaly from around 1824 to 1834 is thought to relate to a forest fire (section 5.6.3).

Whilst not mentioned on Table 5.3, Al is of interest in that it has been shown to increase in the annual rings of *Pinus* in response to increased environmental acidity [63, 84]. Figure 5.18 shows the pattern for Al in Trees C and D. As for Pb concentrations (also shown to increase in *Pinus*) with increasing environmental acidity [151], a gradual rise can be observed from the early 1800's, increasing more steeply at the start of the 1870's and rising to a peak from 1904, peaking in 1911 and 1925 and then declining from 1926 down to a corresponding low in 1934. From here the correlation is not quite so good with Al in Tree C rising towards the very end of the sequence. However the fact that Tree D shows a more pronounced decline suggests that some site specific factor is effecting Tree C from around 1981 onwards. It should be noted that there is no particularly noticeable impact on the Al concentrations in Tree D following fertilisation in the 1960s. The lowest concentration in the sequence is in 1934 for not only Pb and Al, but also Zn, Pr and Nd. This may be explained by the break in SO_4 deposition noted in the Greenland ice cores during the 1930's [116], or maybe connected to increased precipitation for that year diluting down the acidity levels in the soil. It should be noted however that there are higher or very similar precipitation levels at other points in the sequence (for both the yearly and growth season averages), which do not register the same type of effect, so in terms of precipitation it was not an exceptional year.

Finally, Cu appears to show no correspondence with known pollution histories, instead appearing to be dominated by tree physiology (see

CHAPTER 5: CASE STUDY 1 - SWEDEN

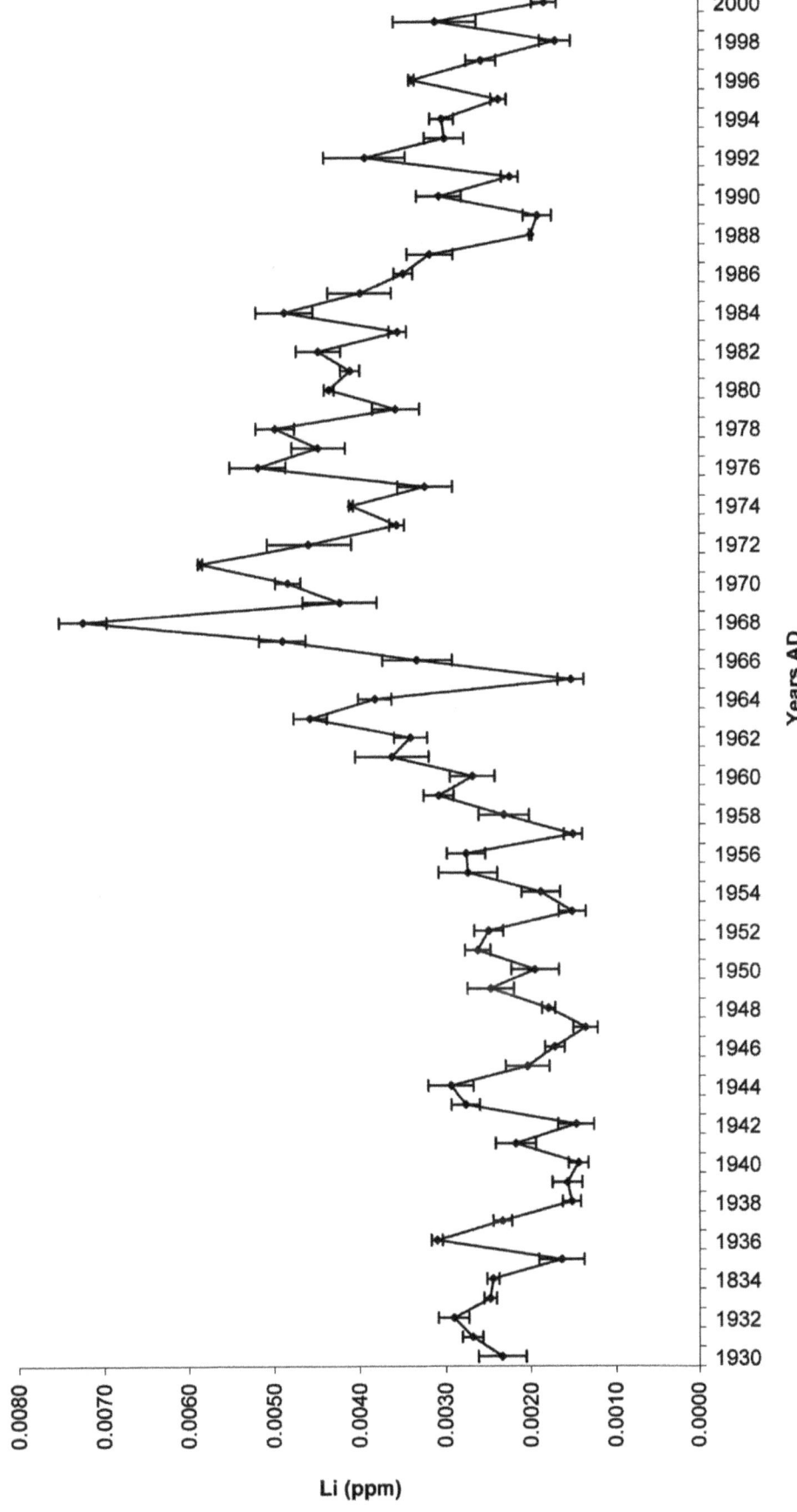

FIGURE 5.15: Cs in Tree D showing an increase in response to fertilisation in 1967

CHAPTER 5: CASE STUDY 1 - SWEDEN

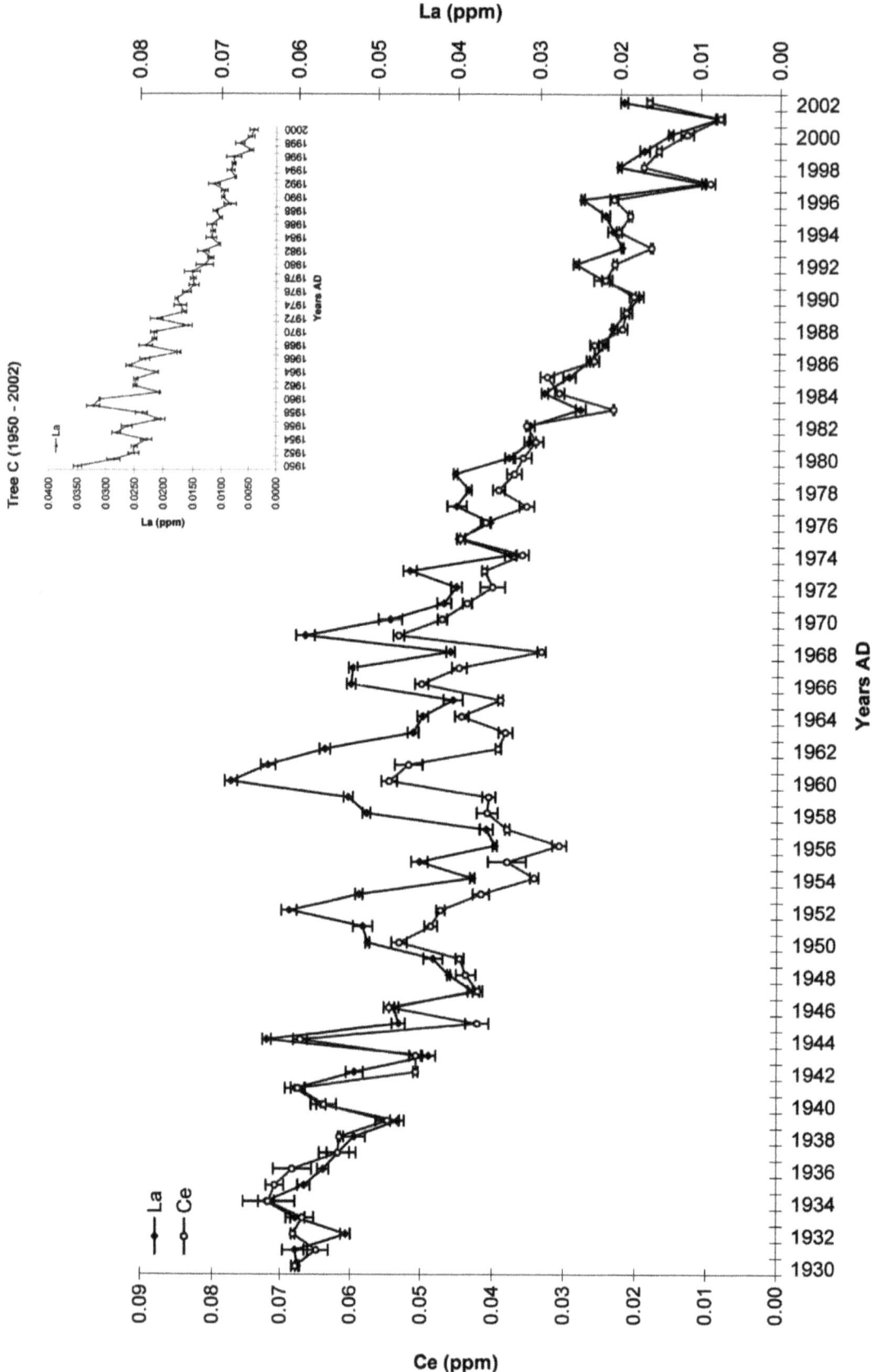

FIGURE 5.16: La and Ce in Tree D (1930 to 2002). Inset shows La in Tree C (1950 to 2002)

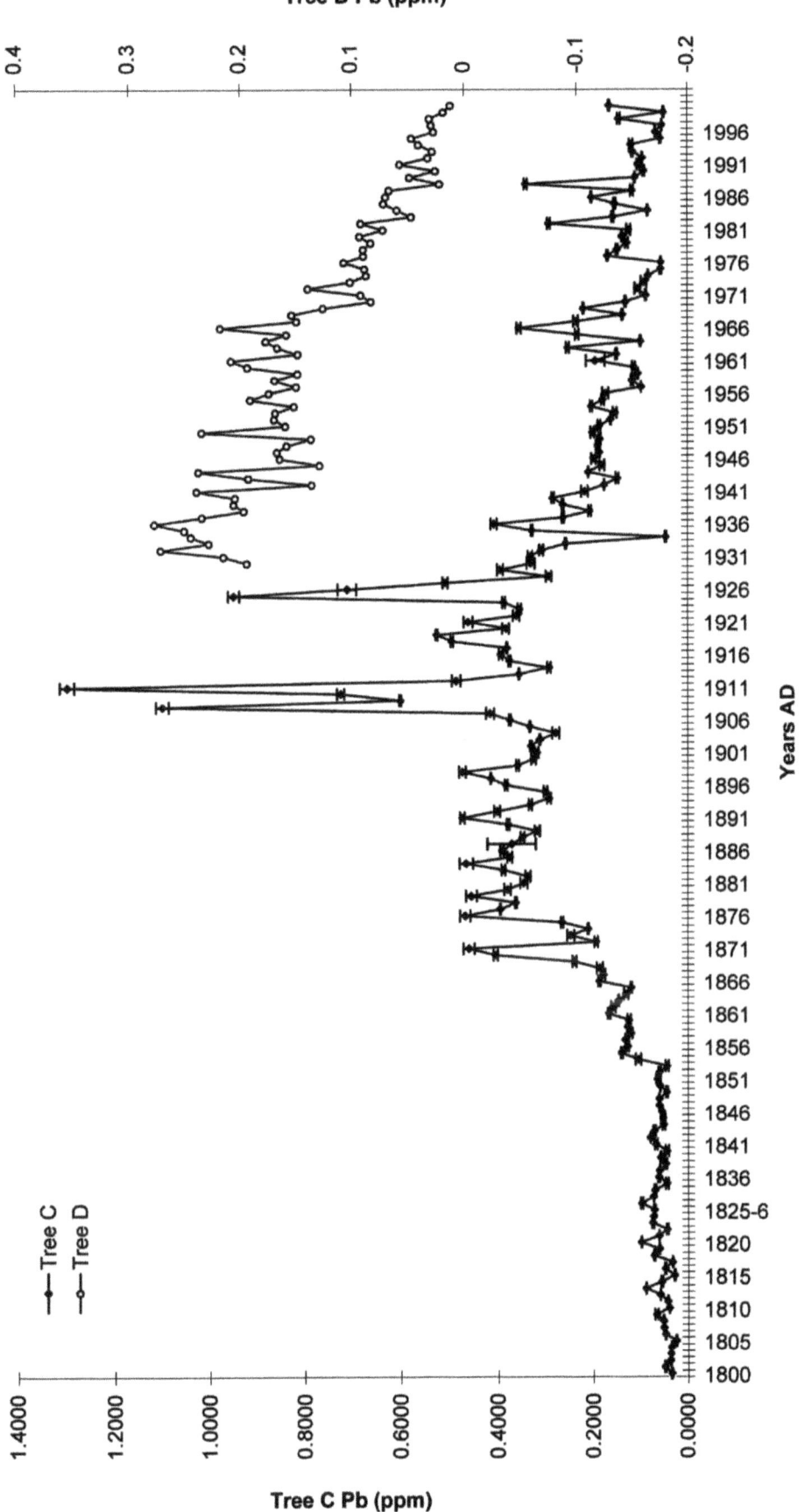

FIGURE 5.17: Pb in Trees C (1800 to 2002) and D (1930 to 2002)

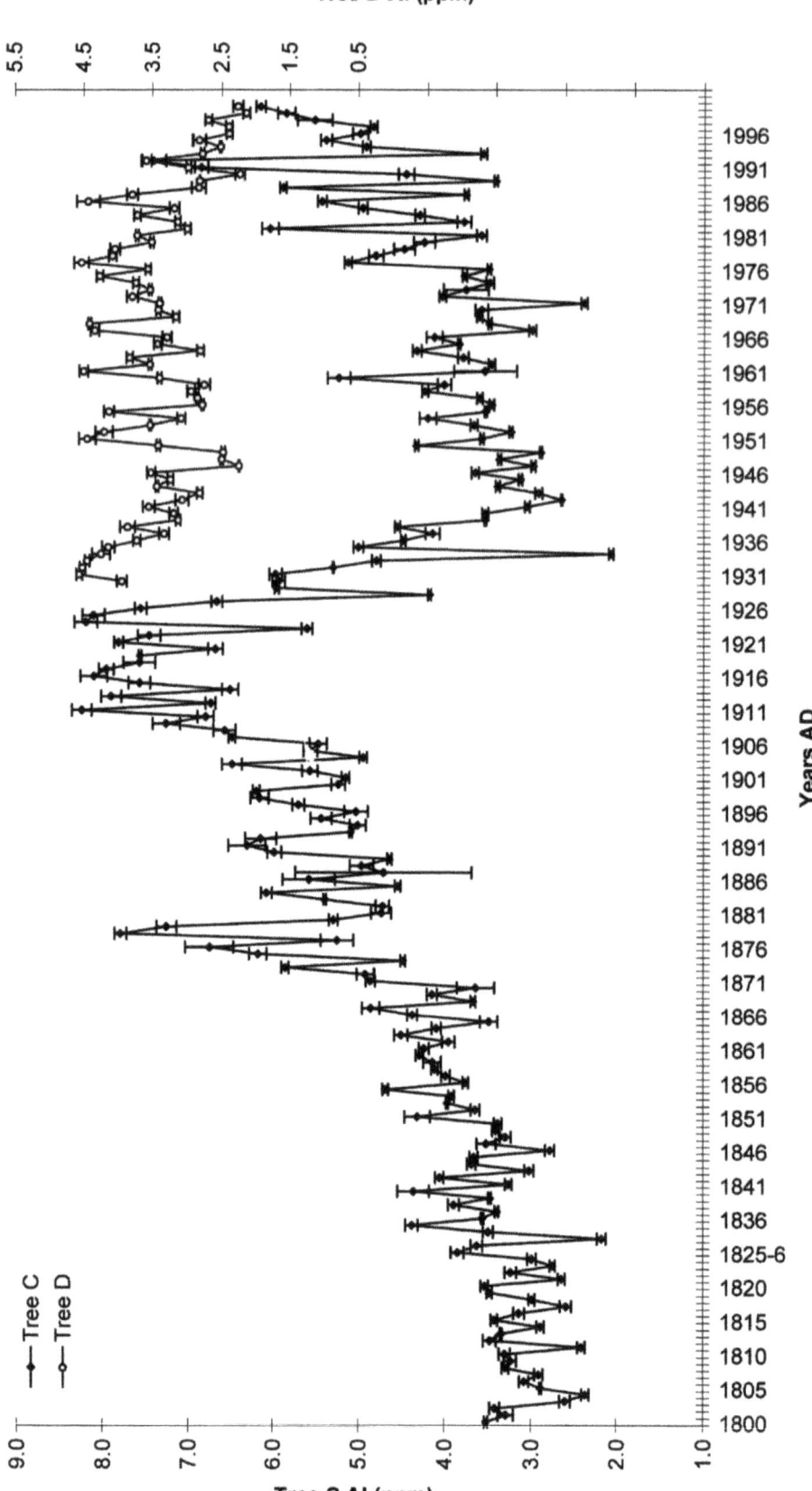

FIGURE 5.18: Al in Trees C (1800 to 2002) and D (1930 to 2002)

section 5.6.1 and figure 5.5). The huge peak in concentrations in 1862 is sufficiently large to be due in all probability, to some kind of contamination during sample preparation or analysis.

Sequences and volcanic eruptions - Whilst there are many undulations in the concentrations of various elements which coincide with the dates of continental and global volcanic eruptions, none would be statistically valid. No signatures are sufficiently clear cut against the noisy background to be truly significant. The only two possible dates for which a tenable association exists are 1875 and 1991 / 1992. The 1875 eruption of Askja resulted in light tephra deposition over central Sweden, therefore it could be argued that of all the volcanic eruptions noted in tables 5.4 and 5.5 it might prove the most likely to result in some form of chemical signature. Figures 5.12, 5.13, 5.14 and 5.18 show varying examples of a short-lived rise in concentrations around this time. However for the majority of elements the rise (with error) is no more substantial than many others in the same sequence, and can be more credibly related to a forest fire around this time (see 5.6.3). The other possible association is in 1991 / 1992. 1991 saw two major volcanic eruptions (Mt. Pinatubo and Cerro Hudson), and several Icelandic events. At this time, a variety of elements (Mn and Cr - 1991, Al, Mn, Rb, Sr, Cd, Sb, Ba, La, Pr, Nd, Cr and Mn - 1992) were observed to display a small peak in concentrations (for examples, see figures 5.12 and 5.18. Whilst there are other, bigger, more noticeable peaks in most of the elemental sequences, this one is shown to be significant by its presence in both trees for such a variety of elements.

Overall, no conclusive evidence was found to support Icelandic volcanism having had any impact on tree ring chemistry in Sweden. Whilst some higher concentrations occur at the time of major volcanic eruptions in one tree they do not in the other, and equally large anomalies occur at other points in the same sequence for which there is no record of volcanism. The lack of a conclusive anomaly for the 1875 Askja eruption, the one event which might have been anticipated to stand out, illustrates the complexity of teasing out the true origin of various elemental traces in tree ring sequences. On a more global scale, if the increase in 1991 / 1992 is connected with an volcanically induced increase in environmental acidity, freeing elements in the soil, or contributing certain elements to the atmosphere for increased uptake by trees, it seems odd that the 1815 eruption of Tambora, known to have released six times more sulphur than Pinatubo [112] does not seem to register any effect.

5.7 Conclusions

Conclusions meeting the specified aims (see section 5.1.1) of the case study are summarised in the next three subsections and in figure 5.19.

5.7.1 Practical procedures

All steps of the sampling methodology were completed successfully with good results. Further improvements for future work would be:

1: further investigation of site specifics in order to choose the best possible site for analysis. In this case the site was chosen because it had a poor soil and low pollutant levels, i.e. to minimise the background noise against which signatures for volcanism might be detected. What the analysis revealed was that even at such a site the tree rings contained a level of background noise which made replication between trees difficult for the majority of elements. It may be that the likelihood of detecting volcanic signatures would be increased in more remote, arid, marginal environments.

2: Increasing the quantity of each tree ring sampled in an attempt to increase the average of each element and so lower the average percentage standard deviation in a single tree ring.

3: To improve aspects of the digestion procedure so that total digestion is achieved removing the need for a filtration step (this would depend upon obtaining a microwave digestion system).

In addition to this, a number of decisions with specific regard to this case study should be discussed. Longer sequences of precipitation and temperature data would have been desirable in order to produce more statistically sound results for the whole sequence of tree rings sam-

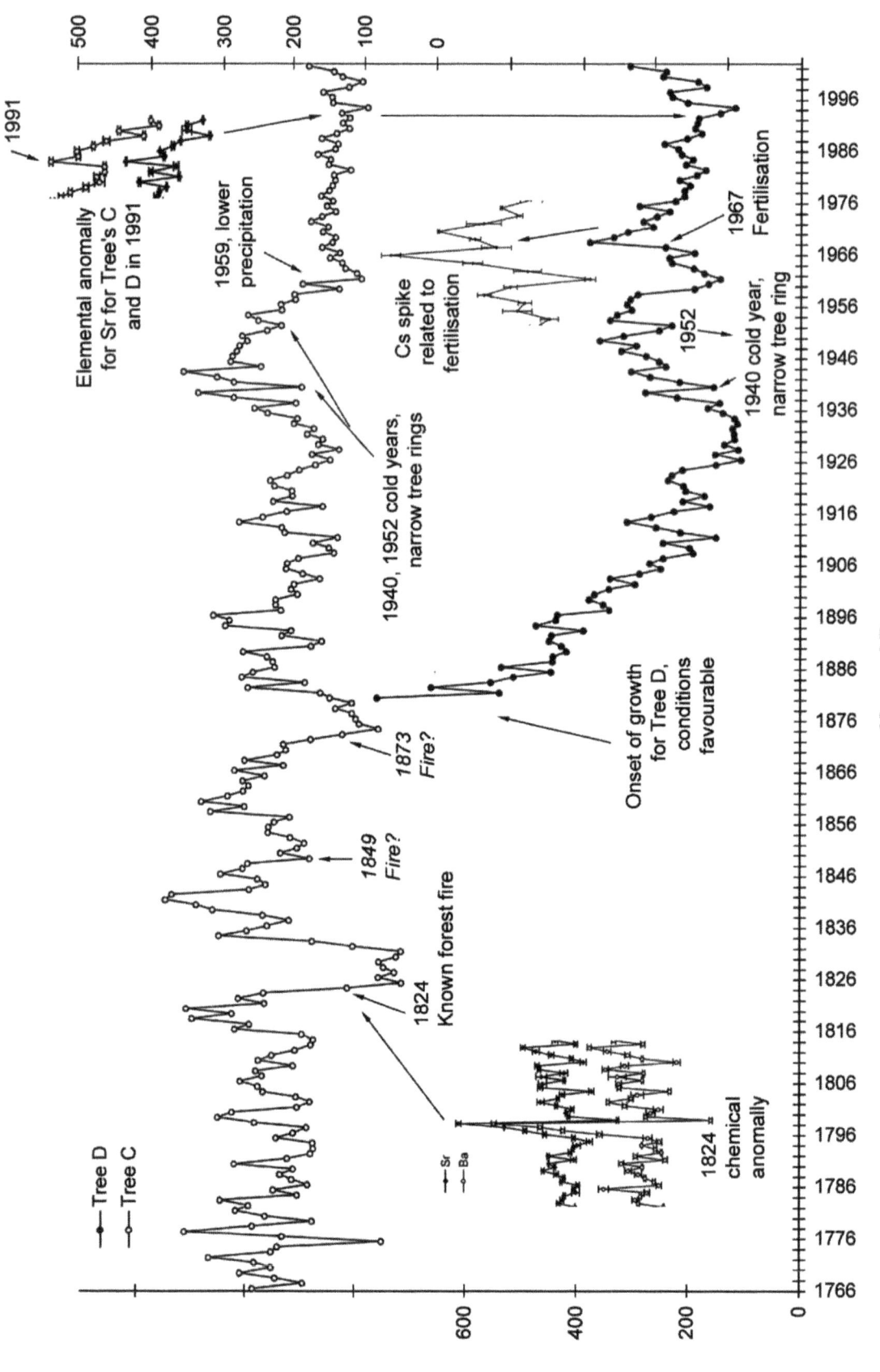

FIGURE 5.19: A summary diagram displaying the main highlights of the data in relation to the life of the two trees.

pled. However, in terms of data specific to the Siljanfors site these were the longest sets available. Whilst it would have been possible to obtain longer records of more general data, it was decided that local data was likely to produce more accurate results and the short sequences were sufficient to gain a thorough understanding of the relationship between these factors and the tree ring sequences. Equally it would have been more desirable to have analysed the same number of tree rings for each of the two trees sampled, and to have added a third sequence from another tree. However, due to the lengthy preparation procedure required for analysis by solution induction ICP-MS, and the time restraints of the project, this was not possible. The necessity to analyse a long sequence of tree rings covering the major volcanic events of the last two hundred years and part of the same sequence from at least one other tree from the same site by way of comparison was given priority. Tree D was shorter lived than Tree C, but the fact that it allowed the opportunity for examining the impact of a deliberate fertilisation event made it a particularly useful sample to select. The third priority was to investigate variation of elemental patterns within the same rings at different heights in a single tree. Again, time limitations meant that a less than desirable number of replications could be made. The final selection of samples for analysis was designed to gain the widest range of suitable data to begin to explore the initial aims and objectives of the case study, this has been achieved, but as ever could be improved by further replication of results.

5.7.2 Endogenous controls

The investigation of the physiological controls on tree ring chemistry, produced some important findings with regard to the use of *Pinus sylvestris* in future dendrochemical studies. The average degree of variation for all elements in one ring was found to be c.60%, showing that even in solutions of the tree rings (which average a sample of around 10000 cells) a high degree of heterogeneity is displayed. This might be improved in future studies by using a larger sample of each tree ring. It was found that the degree of heterogeneity whilst high, does not vary greatly from ring-to-ring, even at different heights in the trunk, i.e. the tree rings are uniformly heterogeneous. It was also possible to show that certain elements are more replicable than others in the same ring, i.e. the degree of replication ranges widely from element to element in a single tree ring. The ten most replicable elements in the sapwood were identified (in descending order) as Sr, Y, La, Ti, Cd, Ba, Al, Pr, and Rb. The ten most reliable in the heartwood were Sr, Cd, Zn, Al, Y, Ba, La, Co, Pr and Ca. The percentage standard deviations produced for all elements could be used to aid interpretation of sequences to elucidate the degree of noise observed in any given pattern and thereby ascertain which peaks in concentration are true peaks and which are within the normal variation. Increases and decreases in the actual concentrations associated with specific elements, were observed with samples taken from increased heights up the trunk. This finding emphasises the need to extract samples from the same height when using several trees for analysis. A change in concentrations was also observed between the heartwood and sapwood, with overall concentrations higher in the heartwood as might be expected in association with the lignification process. On average, concentrations were very similar for the same element in both trees. Be, V, Cr, Ni, As, Se, Sn, Sb, Ho, Er, Tm, Yb, Lu, Hf, Hg, Bi, Th and U were deemed unsuitable for further analysis as they displayed some of the highest variability within a single tree ring and, correspondingly, when plotted in sequence for both trees, the same years displayed no correlation in any part of the sequence. Physiologically derived influences on element patterns (e.g. pith to bark patterns and changes at the heartwood sapwood boundary) were identified for Li, Ca, Zn, Rb, Ti, Cu, and Rh.

5.7.3 Exogenous inputs

Local - The soil at the selected site was suitable for creating optimal conditions for chemical analysis as it was poor, thin and sandy. The presence of slightly more Al, Fe and Ni in relation to less Ca in the leached horizon confirmed the field identification of a podzol. Results suggest that concentrations of Fe, Sr, Ba, Ca and Cr in the tree rings may be governed more by availability in the soil than by the physiological requirements of the tree. This makes them of special interest in tracing chemical signatures. Results also emphasised the complexity of the soil / tree relationship. No strong relationships were found between element patterns, ring widths and climate. Known dates for forest fires only coincided in part with the ring width pattern produced for Tree C, and with the chemical patterns. The narrow ring event from 1824 to 1833 is clearly the worst in the sequence and fits with the given date for a forest fire which obviously put the tree under considerable stress. However, none of the similar anomalies in the sequence fit with the other given date for a forest fire (1857). It is thought that dating of this forest fire may be incorrect. However it is possible that the other anomalies are reflecting some other type of event. In Tree D the impact of fertilisation clearly alters the ring width pattern, however only one element (Cs) showed any particular responsive increase. Overall the two ring width patterns correlate well.

Wider ranging Results indicating the trees were recording some of the wider ranging patterns of anthropogenic deposition were somewhat inconclusive. No evidence was found to connect the elemental patterns in the trees with deposition from Falun, for example. However, data for a number of elements followed the broad background trend of decreasing pollution since the early 1970's. In addition to this, Pb and Al appear to have been responding to changing levels of environmental acidity, where as Cu and Zn seem to be reflecting tree physiology, i.e. they are not available in sufficiently high qualities in the system to follow any external pattern.

In terms of attempting to identify evidence of continental or global volcanism the picture is even more confused. The only signature given credibility by its presence in both trees for the same year for a wide range of elements, is 1991 / 1992. If this association could be proven it would be the first step in beginning to look for evidence of past volcanism in tree rings, however, the lack of signal for bigger eruptions e.g. Tambora 1815 call the peak into question.

5.8 Summary

This case study explored the many variables involved in completing a dendrochemical study on modern wood samples with a view to tracing volcanic eruptions. The results highlighted many key issues in terms of the care with which dendrochemical studies must be approached and the many possible variables which may effect tree ring chemistry. In terms of methodological development, they show that increased care should be taken over site selection in order to ensure the uptake potential of the study trees will be maximised by on site conditions. However, it was demonstrated that in spite of a lengthy sample preparation procedure, and limited facilities for the digestion of samples, it is possible to sample, prepare and run long sequences of individual tree rings via solution induction ICP-MS. The benefits of this technique over LA-ICP-MS were obvious in terms of the range of elements detected. However, the recurring failure to produce clear replication of the same sequences from the same set of tree rings, let alone to produce similar patterns in two trees from the same site, underlines the fact that even this technique may yet be shown to be hindered by the heterogeneous chemistry of certain tree rings.

TABLE 5.13: Eruptions in Iceland over the last 200 years (complied from [15, 142, 145, 155])

Year	Month	Eruption	Location	Explosivity	Duration
2000	February	Hekla	South East	?	c. 3 days
1998	December	Grímsvötn	North East	?	10 days
1996	September	Bárðarbunga / Grímsvötn.	North East	VEI 0	c. 5 days
1991	January	Hekla	South	VEI 3	c. 3 months
1991	September	loki-fogrufjoll	North East	VEI 0	?
1986	November	loki-fogrufjoll	North East	VEI 0	c. 3 days

1984	September	Krafla	North East	VEI 0	14 days
1984	August	Grímsvötn.	North East	VEI 0	<1 day
1983	May	Grímsvötn	North East	VEI 2	c. 5 days
1981	November	Krafla	North East	VEI 0	5 days
1981	April	Hekla	South East	VEI 2	7 days
1980	March,October	Krafla	North East	VEI 0	c. 7 months
1980	August	Hekla	South	VEI 3	3 days
1977	April	Krafla	North East	VEI 0	c. 5 months
1975	December	Krafla	North East	VEI 1	1 day
1973	January	Heimaey	South	?	?
1972	March	Grímsvötn	North East	VEI 0	1 month
1970	May	Hekla	South	VEI 3	c. 2 months
1968	May	Kverkfjöll	East	VEI 0	c. 6 days
1961	November	Askja	North East	VEI 2	c. 2 months
1959	Unknown	Kverkfjöll	East	VEI 0	?
1955	June	Katla	South	VEI 1	1 day
1954	July	Grímsvötn	North East	VEI 2	1 day
1948	February	Grímsvötn	North East	VEI 0	1 day
1947	March	Hekla	South	VEI 4	c. 13 months
1941	April	Grímsvötn	North East	VEI 0	4 months
1939	June	Grímsvötn	North East	VEI 0	1 day
1938	December	Askja	North East	VEI 2	?
1938	May	Grímsvötn	North East	VEI 1	1 day
1934	March	Grímsvötn	North East	VEI 2	c. 9 days
1934	December	Grímsvötn	North East	VEI 2	c. 5 days
1933	November	Grímsvötn	North East	VEI 1	c. 10 days
1929	January	Kverkfjöll	East	VEI 1	c. 1 month
1927	September	Esjufjöll	East	VEI 1	?
1926	June	Reykjaneshryggur	West	VEI 0	?
1926	July	Askja	North East	VEI 2	?
1924	?	Askja	North East	VEI 0	?
1923	January	Askja	North East	VEI 0	?
1922	November	Askja	North East	VEI 0	?
1922	September	Grímsvötn	North East	VEI 2	c. 25 days
1921	March	Askja	North East	VEI 0	?
1919	Unknown	Askja	North East	VEI 2	c. 1 day
1918	October	Katla	South	VEI 4	c. 23 days
1913	April	Hekla	South	VEI 2	c. 1 month
1910	June	Loki-Fogrufjoll	North East	VEI 2	c. 4 months
1902	December	Bárðarbunga	North East	VEI 2	c. 6 months
1891	November	Grímsvötn	North East	VEI 2	c. 4 months
1883	January	Grímsvötn	North East	VEI 2	3 months
1878	February	Hekla	South	VEI 2	c. 2 months
1875	March	Askja	North East	VEI 4	c. 2 days
1873	January	Grímsvötn	North East	VEI 4	c. 7 months
1872	Unknown	Bárðarbunga	North East	VEI ?	?
1867	August	Grímsvötn	North East	VEI 1	1 day
1862	June	Bárðarbunga	North East	VEI 2	c. 28 months
1861	May	Grímsvötn	North East	VEI 2	?
1860	May	Katla	South	VEI 3	c. 19 days
1854	?	Grímsvötn	North East	VEI 2	?
1845	September	Hekla	South	VEI 4	c. 7 months
1838	June	Grímsvötn	North East	VEI 2	c. 1 day
1823	July	Katla	South	VEI 3	c. 1 month
1823	February	Grímsvötn	North East	VEI 2	c. 1 day
1821	December	Eyjafjoll	South	VEI 2	13 days

| 1816 | May | Grímsvötn | North East | VEI 2 | c. 1 month |
| 1807 | ? | Bárðarbunga | North East | VEI 0 | ? |

Chapter 6

Case Study 2 - Porsuk, Turkey

6.1 Introduction

As a major end goal of this research project is to trace the volcanic eruptions of prehistory in tree rings which, given their age, will have come from some sort of environment of preservation, it was important to assess the potential of dendrochemical technique specifically in this context. The selected samples for this pilot study are especially relevant to the overall investigation, as they include a growth anomaly thought to relate to the environmental impact of the Bronze Age eruption of Thera on the Aegean archipelago of Santorini [96]. This was a major volcanic event in an area of high civilization. Its immediate effect was the burial of Bronze Age Akrotiri, but the wider impact has been the subject of great dispute for over sixty years [9, 11, 23, 24, 53, 60, 61, 62, 68, 79, 94, 95, 96, 123, 169, 172, 173]. An absolute date for Thera would provide a key starting point for the analysis and interpretation of the wide ranging impact of this dramatic event. However, even more critically, if an absolute date could be attributed to the associated tephra horizon, which spans the entire Aegean region and beyond, it would provide a pivotal, absolute date which could be used to calibrate all other chronologies of the region, including the historical calendar chronologies of Egypt and Mesopotamia.

The two samples, C-Tu-Por26A and C-Tu-Por3A (see figure 6.1) were obtained via the Aegean Dendrochronology Project, from an excavated structure in Porsuk, Turkey. They were charred timbers of Middle Bronze Age date which had been re-used in a Hittite rampart. Further information was requested regarding the chemistry of the sediment in which the samples had been embedded, how ever none was available. It is speculated that the trees from which the samples came initially grew in the Taurus mountain range to the south of the site (Peter I. Kuniholm, pers. comm. 2003). The site itself is situated approximately 840 kilometres downwind of Thera. Both samples were dated as part of the Aegean chronology and cover the exceptional growth ring for relative year 854 - dated 1650 BC +4 / -7 [96], which may perhaps correlate with the large volcanic signal noted by Hammer et al. [61] and Clausen et al. [29] and which in turn is thought to represent the Thera eruption [29, 61, 95, 96]. Direct radiocarbon evidence at present places the Thera eruption within the period 1664-1615 BC at 95.4% confidence (Manning pers. comm.).

6.2 Aims

There were three main aims to this case study. First, to conduct chemical analysis of individual tree rings from a sample of archaeological material likely to cover one of the major volcanic eruptions of prehistory. Second, to use three complementary analytical techniques in this context in order to maximize and verify findings. Third, to ascertain if any chemical signature could be found in the tree rings to directly link the eruption of Thera with the ring 854 growth anomaly.

Figure 6.1: Samples (A) C-Tu-Por 3A and (B) C-Tu-Por 26A

6.3 Laboratory preparation

Samples were received pre-measured from Cornell University, however several radii were remeasured (with the addition of extra pin marks) at Reading prior to cleaning and dissection. This was done so that if the samples were to fragment on beginning dissection, (as was anticipated owing to the friable nature of the semi-carbonised wood), it would still be possible to identify all sections by their relative years. The samples were cleaned by physical removal of the outer surface using a steel razor blade. The presence of root material and minute clasts of the wall matrix were noted, however there was insufficient quantity of the matrix to perform a separate chemical analysis. Sub-samples of complete radii were sectioned for each sample and mounted for laser ablation sampling. This was easier to achieve with C-Tu-Por3A than C-Tu-Por26A as C-Tu-Por26A was far more friable. Individual tree rings for solution analysis were dissected from the main samples with steel razor blades. The samples were lightly ground in an acid washed, agate pestle and mortar in order to homogenize and reduce the particle size for more effective digestion. Material from the cleaning procedure was used to carry out experiments to ascertain the optimum time required for ashing and subsequently digesting the samples. It was found that ashing at 400°C for nine hours with venting every three hours to introduce more oxygen, produced the best results. Whole sequences were prepared at the same time and then cold stored until analysis.

The three analytical techniques selected for comparative analysis were ICP-AES, ICP-MS and LA-ICP-MS. ICP-AES was selected primarily to analyse for possible volcanically derived S (which does not ionise and so cannot be detected by mass spectrometry techniques). Ag, Al, Ba, Ca, Cu, Fe, Mg, Mn, Sr, Zr, and Zn were also analysed via this technique in order to provide a range of element concentrations to compare with the other techniques. ICP-MS was selected as the method of choice developed as part of the overall research project, with the beneficial potential of none-destructive LA-ICP-MS being tested in this, the context of a rare archaeological sample. The following sample batches were prepared and analysed:

1. C-Tu-Por26A (years 840-870) [1] solutions for ICP-AES analysis (0.2g original sample, 0.5ml HNO_3, in 20mls H_2O)

2. C-Tu-Por3A (years 825-882) solutions for ICP-AES analysis (0.2g original sample, 0.5ml HNO_3, in 20mls H_2O)

3. C-Tu-Por26A (years 840-870) solutions for ICP-MS analysis (Sub-sample of 1)

4. C-Tu-Por26A (years 836-860) solutions for ICP-MS analysis (0.2g original sample, 0.5ml HNO_3, in 5mls H_2O)

5. C-Tu-Por26A (years 840-875) laser samples for LA-ICP-MS analysis (solid sample)

6. C-Tu-Por3A (years 814-887) solutions for ICP-MS analysis (0.2g original sample, 0.5ml HNO_3, in 20mls H_2O)

7. C-Tu-Por3A (years 825-894) laser samples for LA-ICP-MS analysis (solid sample)

Due to technical and instrumental failure above and beyond the control of the analyst, data generated for the following sample batches were rejected as insufficiently reliable for interpretation: C-Tu-Por3A (years 814-887) solutions for ICP-MS analysis (technical set up failure, mass calibration drift). C-Tu-Por26A (years 825-894) laser samples for LA-ICP-MS (leaking seal in the instrument on the day of analysis).

6.4 Results

Figures 6.2 to 6.14 show a comparison of concentrations for each of the various elements in the two samples following analysis by ICP-AES.

The two samples were found to display similar concentration levels (see Table 6.1), however 26A had on average higher concentrations than 3A, with the exception of Ba and Zn which were lower. Al, Ag, Cu, Fe and Mn showed no correlation between annual concentrations for the two samples.

---

[1] sample sequence length and start date were determined by the maximum material available for extraction from each sample

Chapter 6: Case Study 2 - Porsuk, Turkey

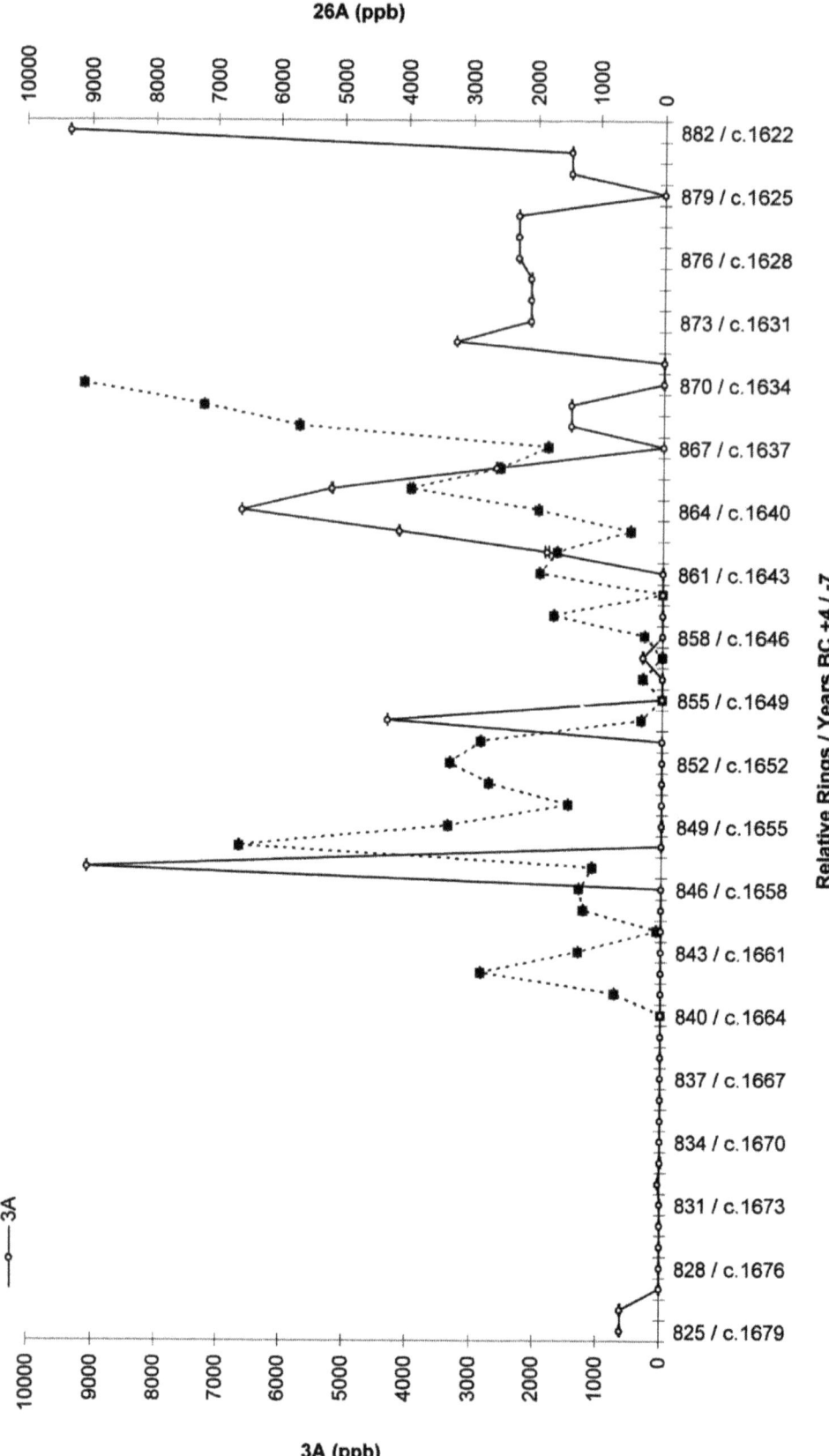

Figure 6.2: Al concentrations (ppb) in Por-Tu-26A and Por-Tu-3A

CHAPTER 6: CASE STUDY 2 - PORSUK, TURKEY

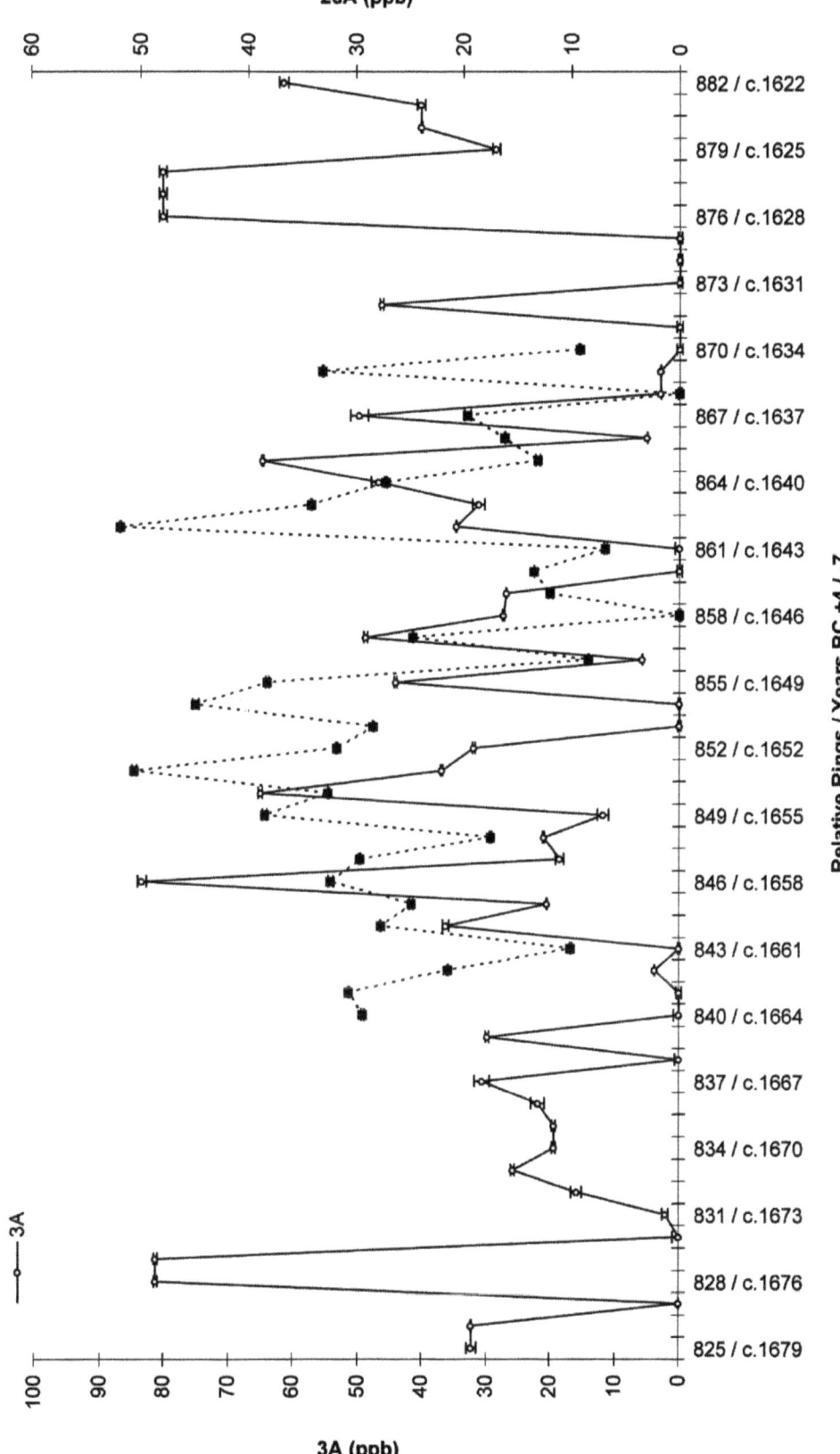

FIGURE 6.3: Ag concentrations (ppb) in Por-Tu-26A and Por-Tu-3A

CHAPTER 6: CASE STUDY 2 - PORSUK, TURKEY

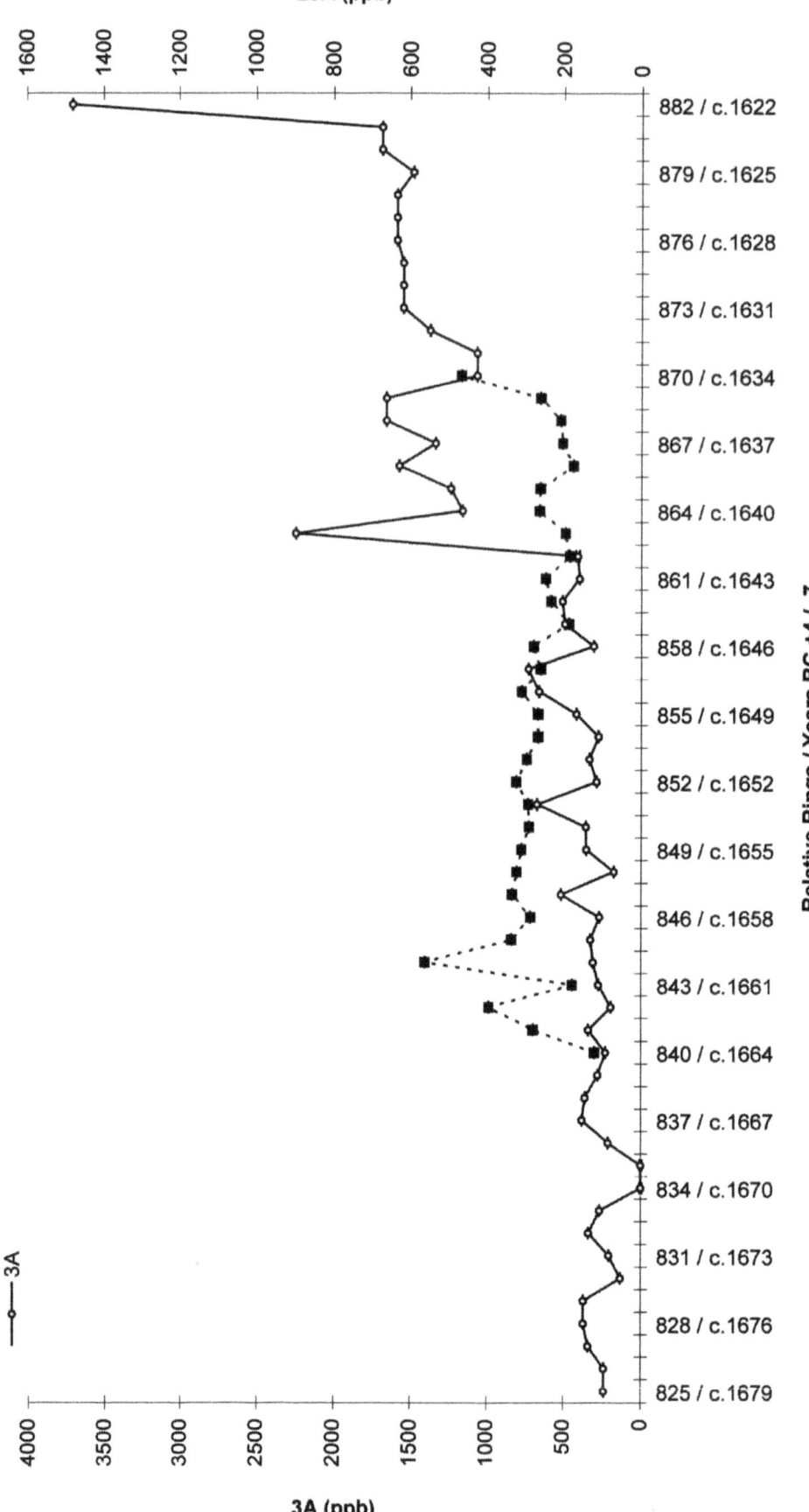

FIGURE 6.4: Cu concentrations (ppb) in Por-Tu-26A and Por-Tu-3A

CHAPTER 6: CASE STUDY 2 - PORSUK, TURKEY

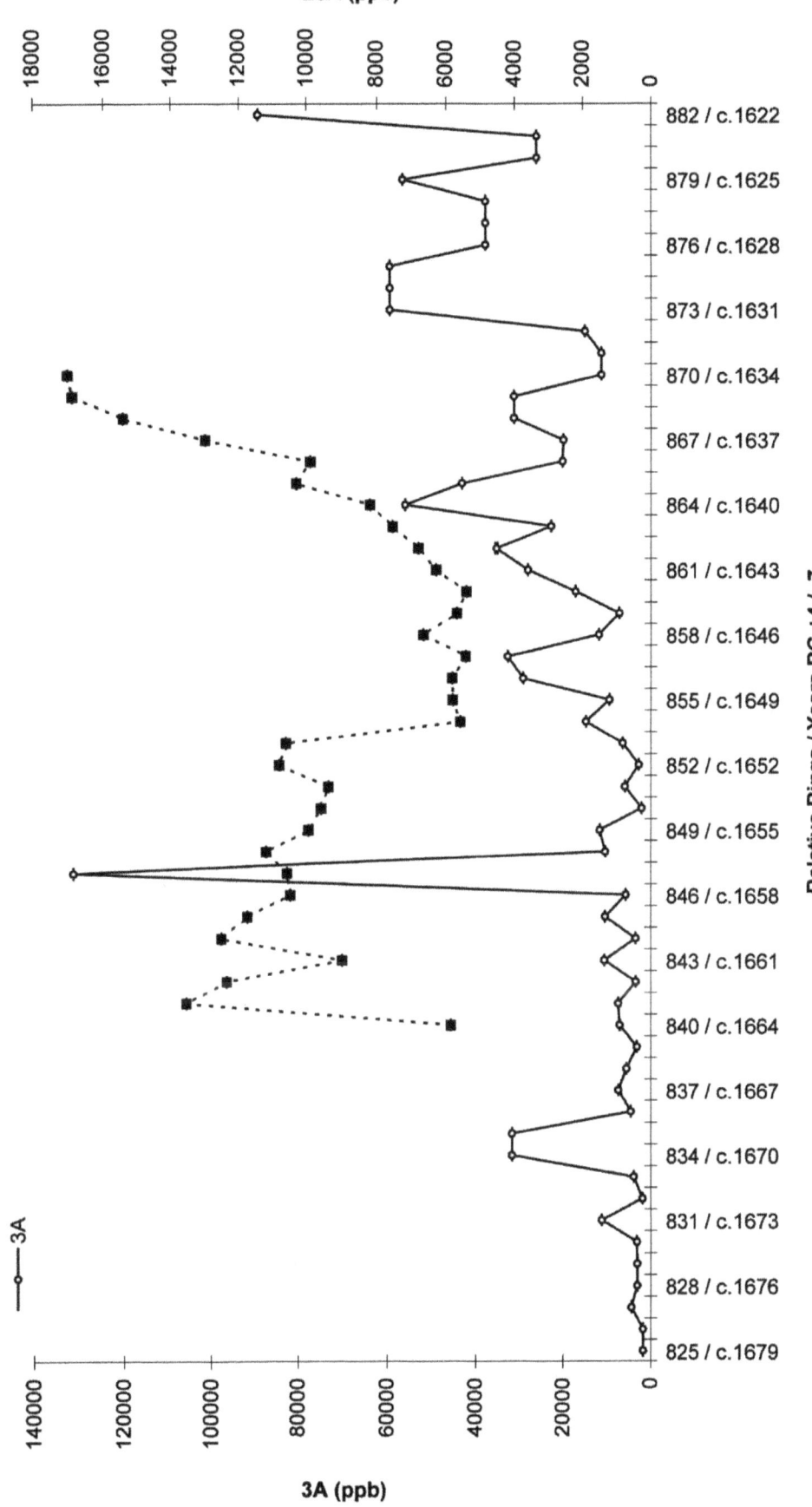

FIGURE 6.5: Fe concentrations (ppb) in Por-Tu-26A and Por-Tu-3A

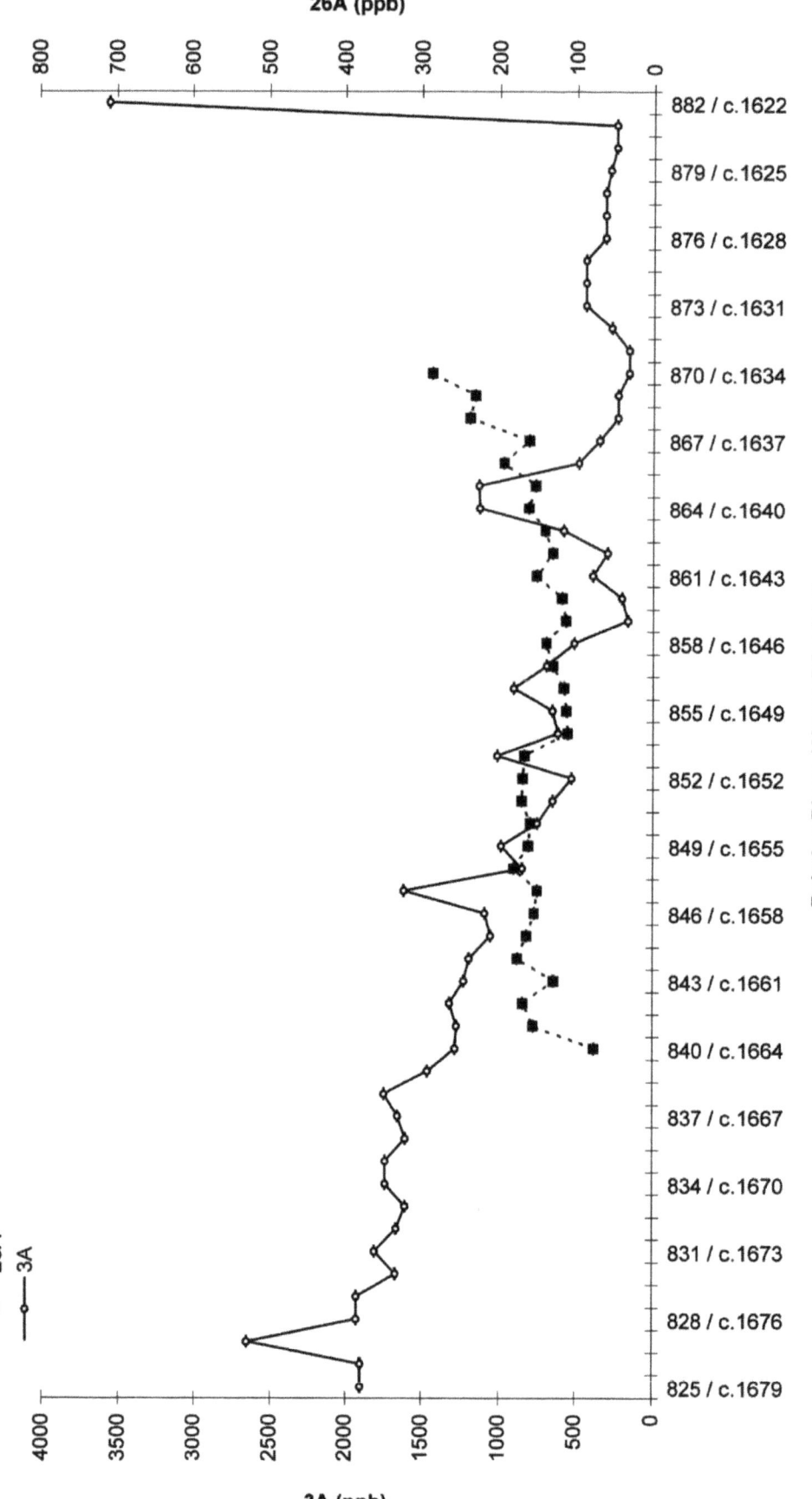

FIGURE 6.6: Mn concentrations (ppb) in Por-Tu-26A and Por-Tu-3A

CHAPTER 6: CASE STUDY 2 - PORSUK, TURKEY

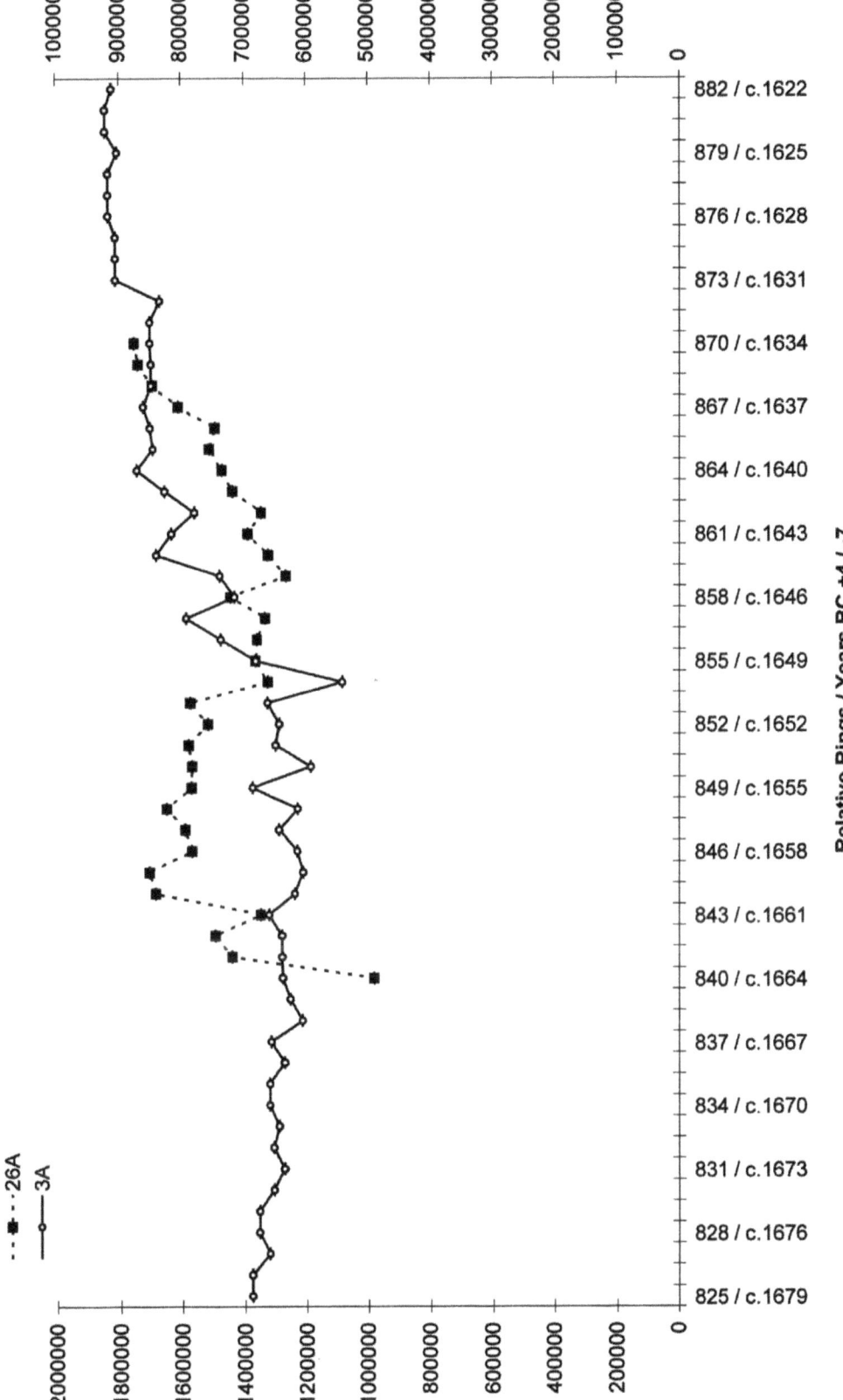

FIGURE 6.7: Mg concentrations (ppb) in Por-Tu-26A and Por-Tu-3A

CHAPTER 6: CASE STUDY 2 - PORSUK, TURKEY

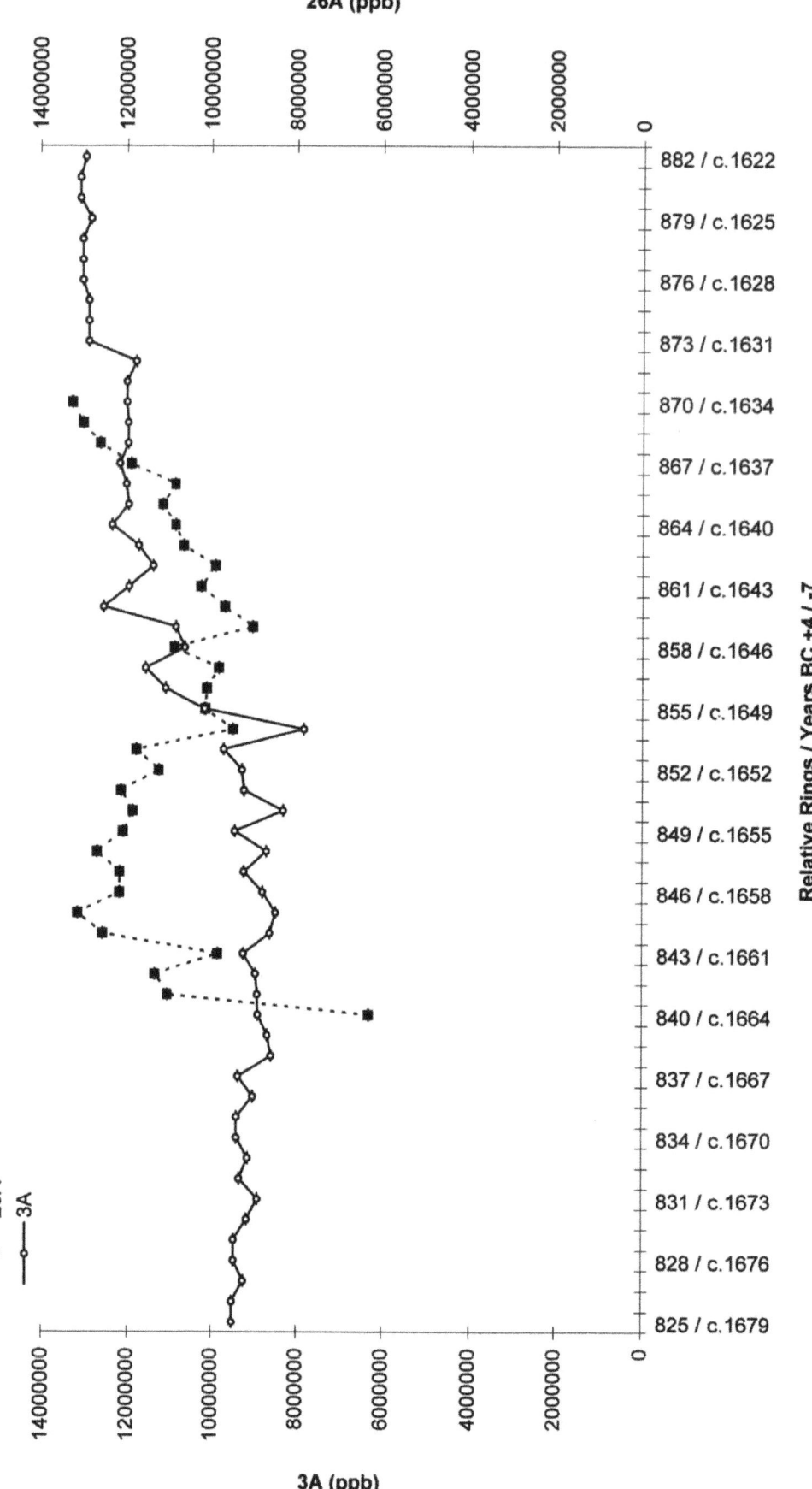

FIGURE 6.8: Ca concentrations (ppb) in Por-Tu-26A and Por-Tu-3A

CHAPTER 6: CASE STUDY 2 - PORSUK, TURKEY

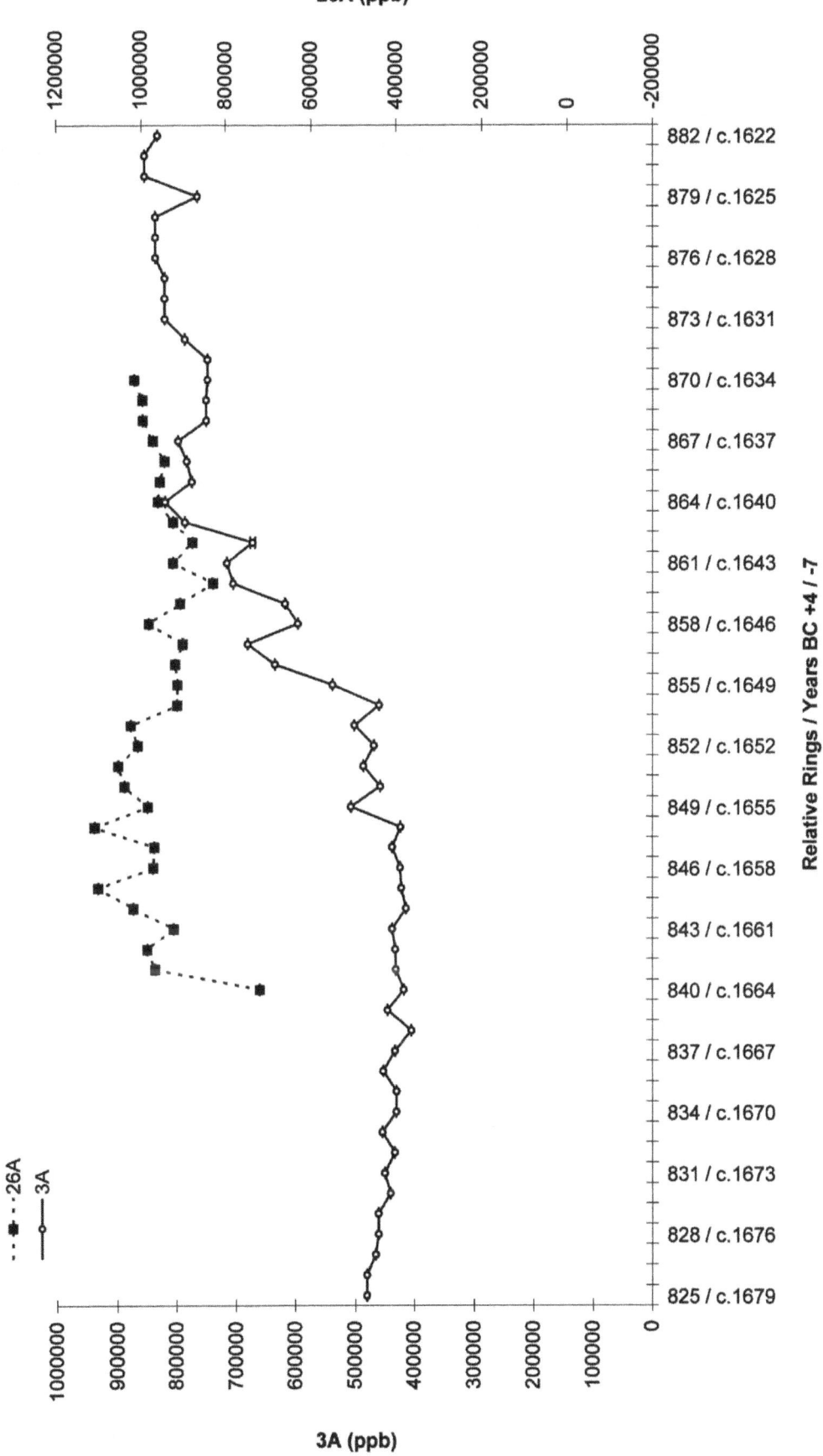

FIGURE 6.9: Na concentrations (ppb) in Por-Tu-26A and Por-Tu-3A

CHAPTER 6: CASE STUDY 2 - PORSUK, TURKEY

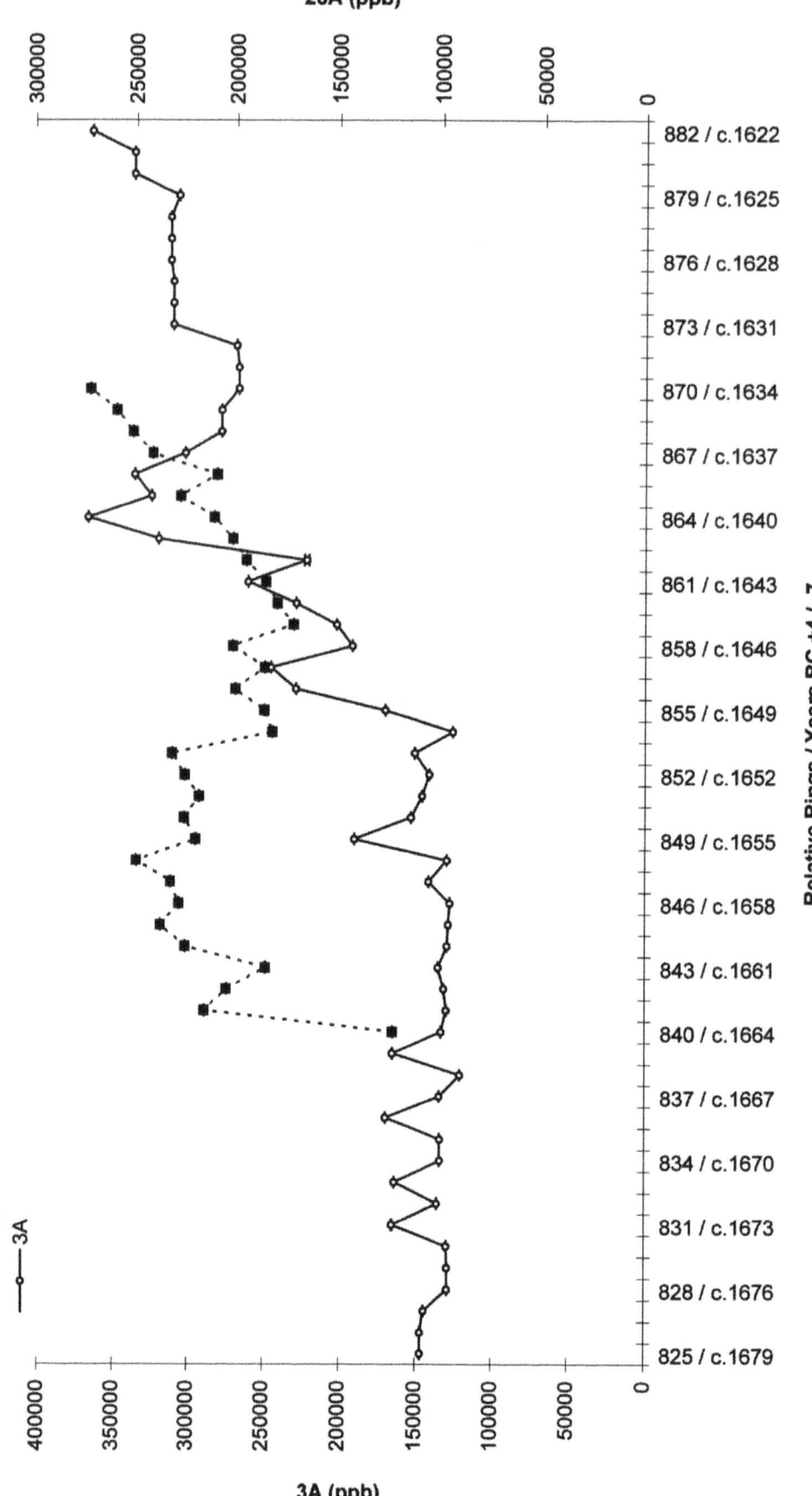

FIGURE 6.10: S concentrations (ppb) in Por-Tu-26A and Por-Tu-3A

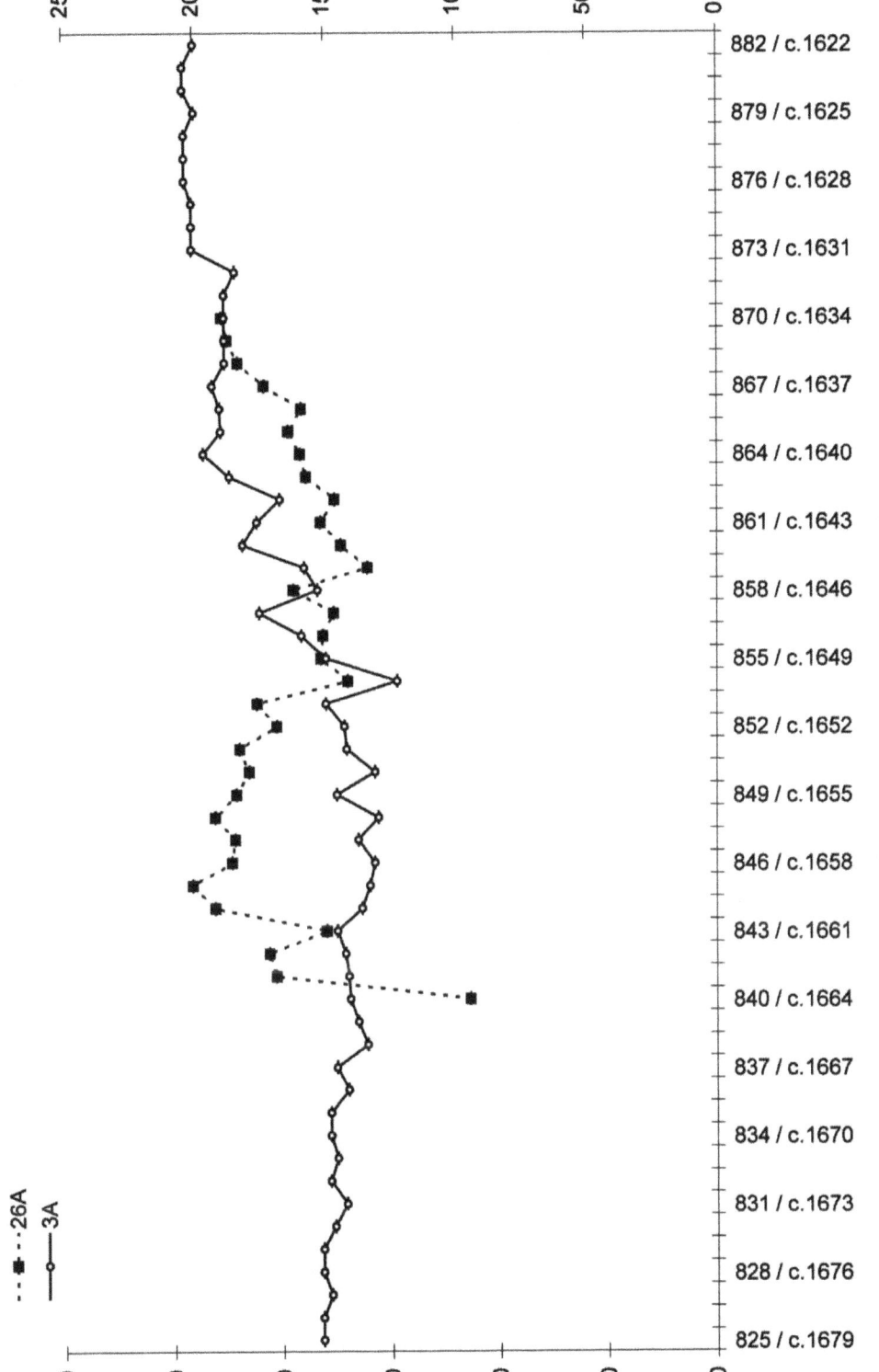

FIGURE 6.11: Sr concentrations (ppb) in Por-Tu-26A and Por-Tu-3A

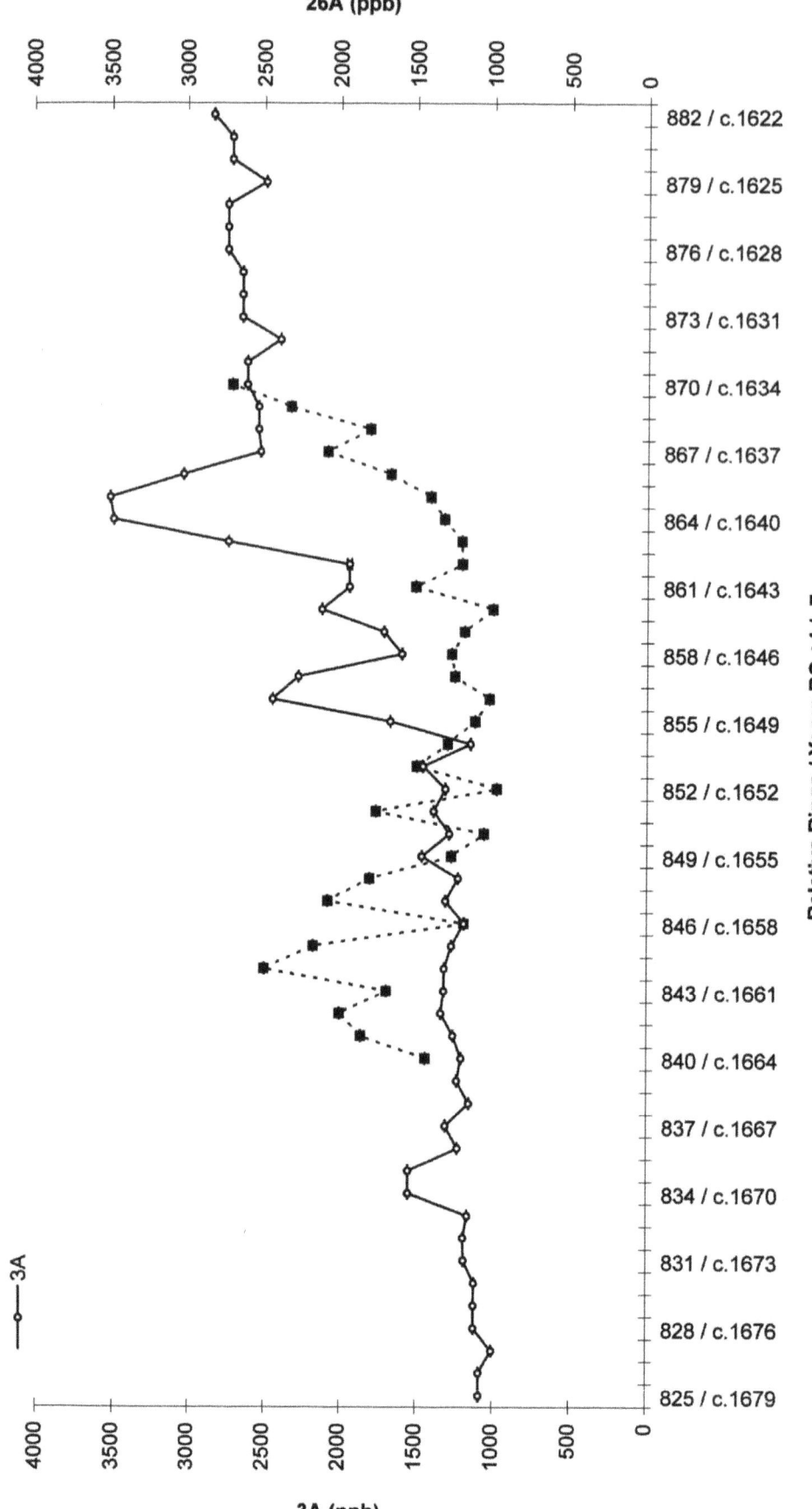

FIGURE 6.12: Ba concentrations (ppb) in Por-Tu-26A and Por-Tu-3A

CHAPTER 6: CASE STUDY 2 - PORSUK, TURKEY

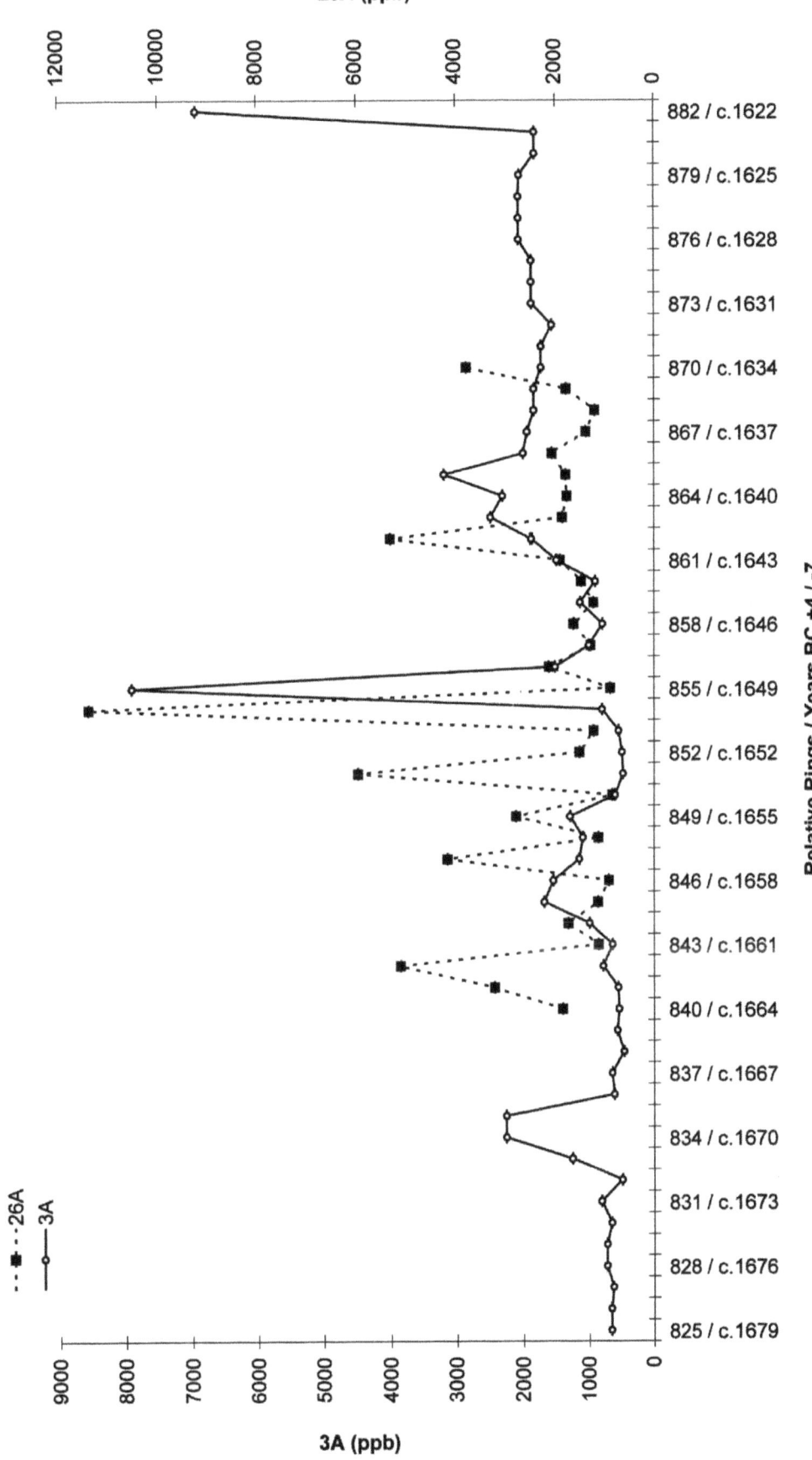

FIGURE 6.13: Zn concentrations (ppb) in Por-Tu-26A and Por-Tu-3A

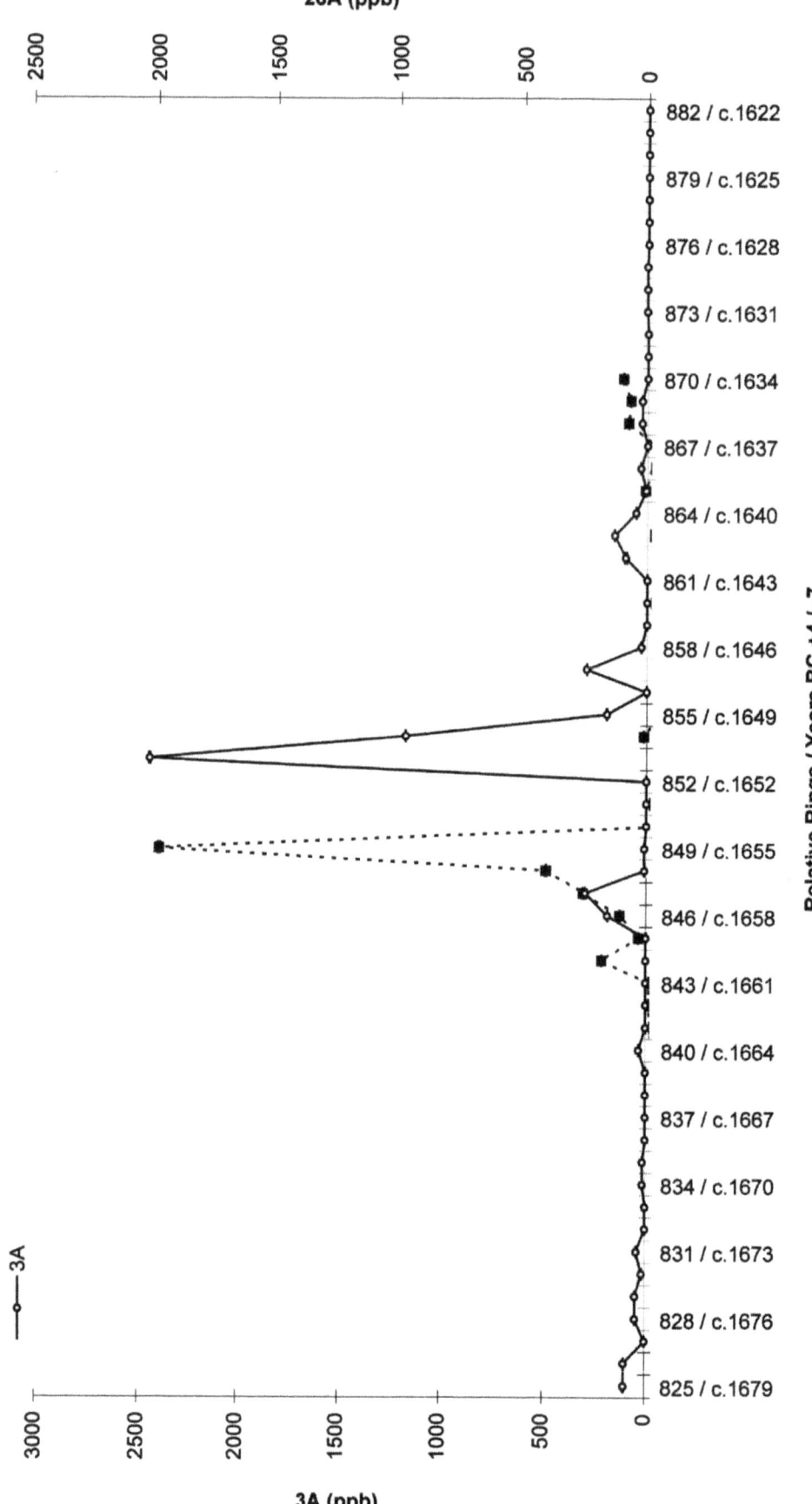

FIGURE 6.14: Zr concentrations (ppb) in Por-Tu-26A and Por-Tu-3A

TABLE 6.1: Average concentrations of elements from samples 3A and 26A via ICP-AES, with a replicate run of the same solutions from 26A via ICP-MS. Values in ppb. Where 'nd', the element was not determined via ICP-MS, Mg was detected but over saturated due to the unusually high concentrations

Element	3A (ICP-AES)	26A (ICP-AES)	26A (ICP-MS)
Ag	16	41	56
Al	1150	2190	4340
Ba	1900	1600	1500
Ca	12000000	11000000	1100000
Cu	310	690	950
Fe	2800	75000	42000
Mg	740000	1500000	-
Mn	200	800	900
Na	630000	830000	nd
S	160000	280000	nd
Sr	160000	200000	190000
Zn	2000	1800	1500
Zr	78	94	45

Mg, Ca, Na, Mn, S, and Sr displayed similar concentration patterns to one another within each tree for both samples, and Ba, Ca, Mg, Na, S, Sr, and Zn displayed slight correlations between samples at the same point in the sequences from both trees. Of particular note are the data for Mg, (figure 6.7) Ca, (figure 6.8) S (figure 6.10 and Sr (figure 6.11). Here for both data sets a depletion can be observed for event year 854 followed by a gradual rise. Whilst this effect is most marked in the data for 3A (also for other elements such as Ba and Fe), the fact that there appears to be partial correlation for certain elements in 26A may be significant, albeit that the correlation is not statistically significant. The Zn spike at 855/856 for 3A and 854 in 26A may also have some significance as it may relate to a sudden influx of this element into the environment around the event year. The spikes displayed for Zr, are also of interest in that the one occurring for 3A matches the Zn spike for this sample and that 26A appears to mirror the spike a few years earlier. Although the Fe pattern for 26A does not match with that of 3A, it does appear to show a negative, followed by positive response around 854.

In order to investigate these data further each element sequence was plotted against the tree ring width pattern for each of the trees. These results provided a graphical representation of the previous observations, e.g. figure 6.15

Triplicate solutions covering years 847 to 861 were prepared separately from the same bulk sample extracted for each year in order to check replication potential for the previously discussed patterns, i.e. to estimate the error introduced by the preparation procedure and / or the heterogeneity of the wood. Results showed that S, Zr, Sr, Na, Fe and Mg were the most reliably replicated concentration patterns with very similar values detected for each of the three separately prepared batches (for example see figure 6.16). Whilst overall Mn, Ca, and Ba also show good replication a number of samples do not correlate so well (see figure 6.17). Al shows good agreement for two of the samples and less for the third. Neither Zn or Cu were replicable, for example, see figure 6.18.

A further test was to check the replication potential for the same sequence of solutions via a different instrumental technique, i.e. to estimate the error introduced by specific analytical techniques. A sub-sample of the ICP-AES solution sequence for 26A was taken and run via ICP-MS. Results showed a good degree of agreement with concentrations for various elements via the two techniques (see table 6.1). The concentrations are of the same orders of magnitude, with differences explained to some extent by differences in the precision of the two techniques, as shown in table 6.2. Although this relationship is not hugely apparent, the overall levels of precision are good and when the data are plotted graphically visual agreement is strong for most elements. See figure 6.19.

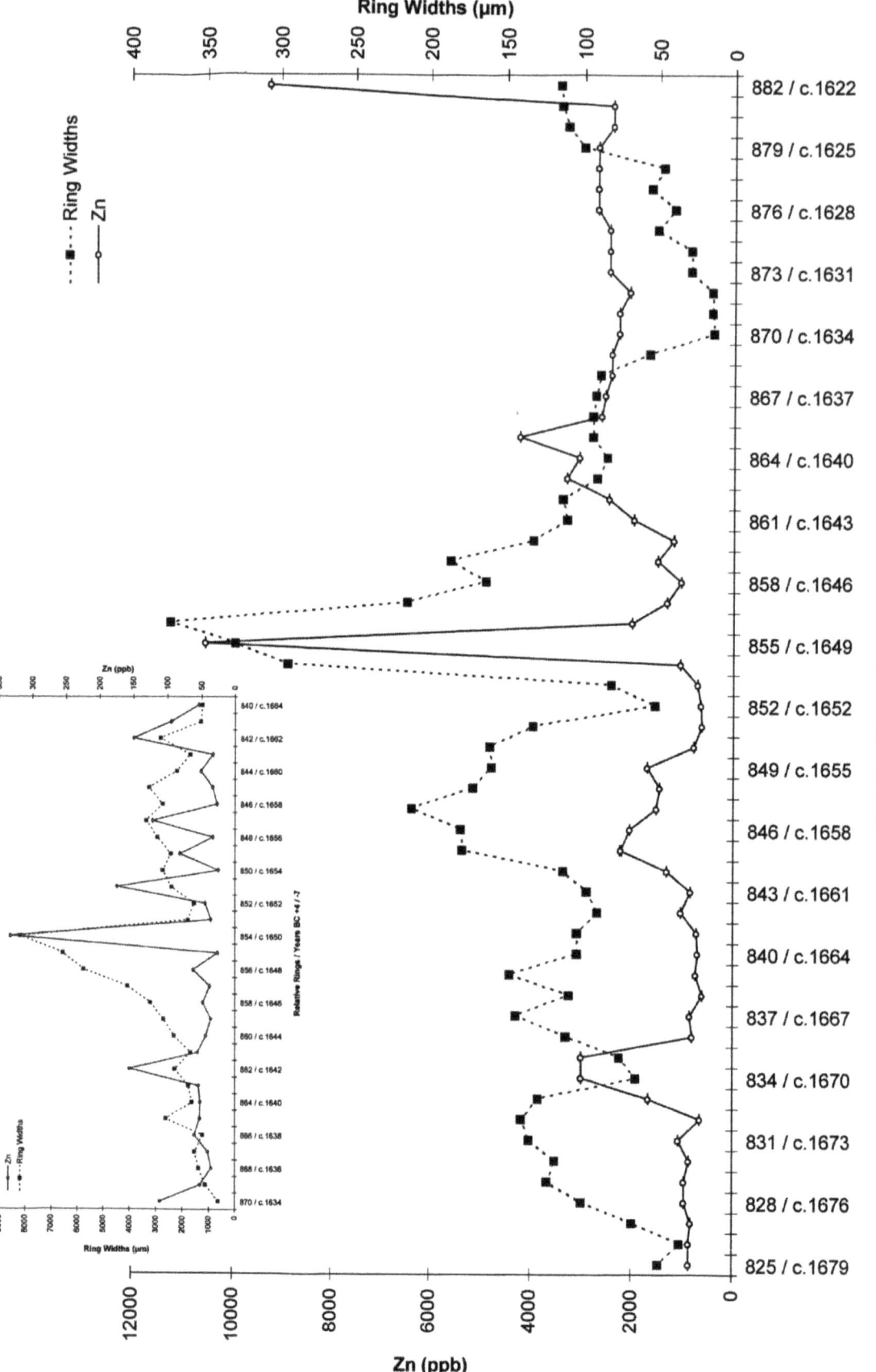

FIGURE 6.15: Zn concentrations (ppb) against ring width pattern for Por-Tu-3A. Inset comparative data for Por-Tu-26A

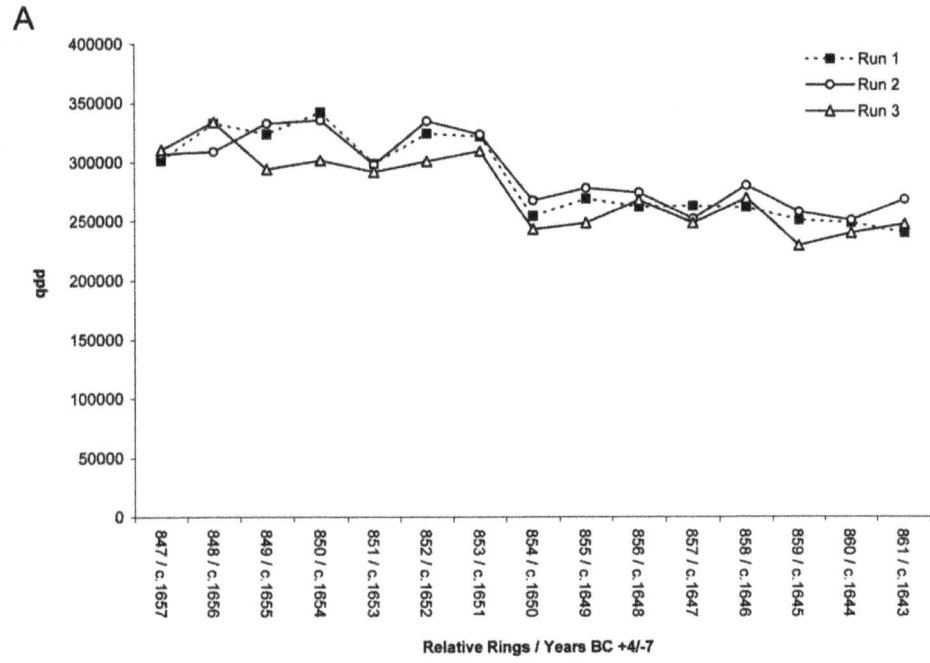

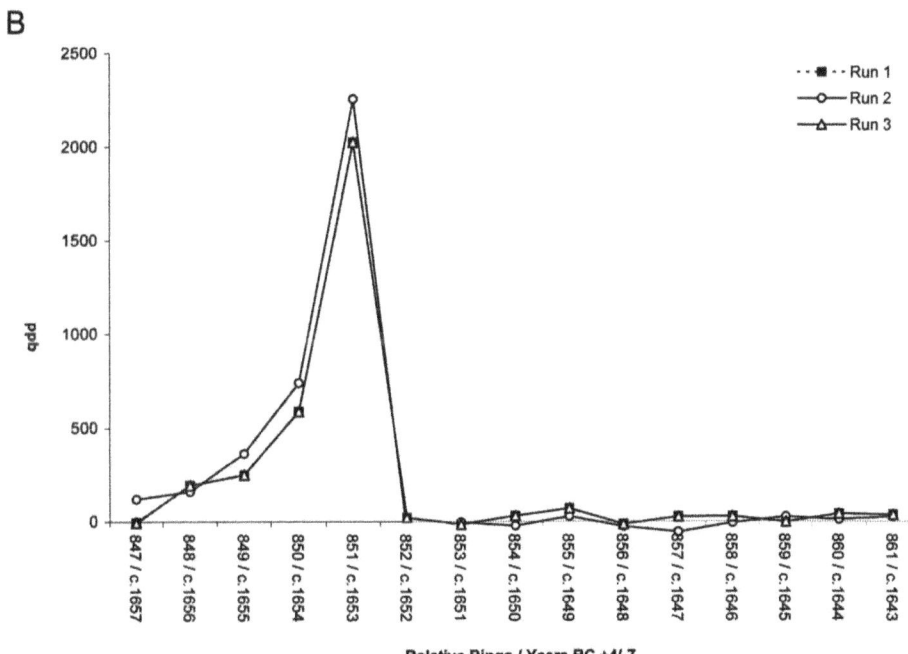

FIGURE 6.16: Example of correlating concentration patterns for S and Zr in three replicate sequences from Por-Tu-26A. A is S, B is Zr

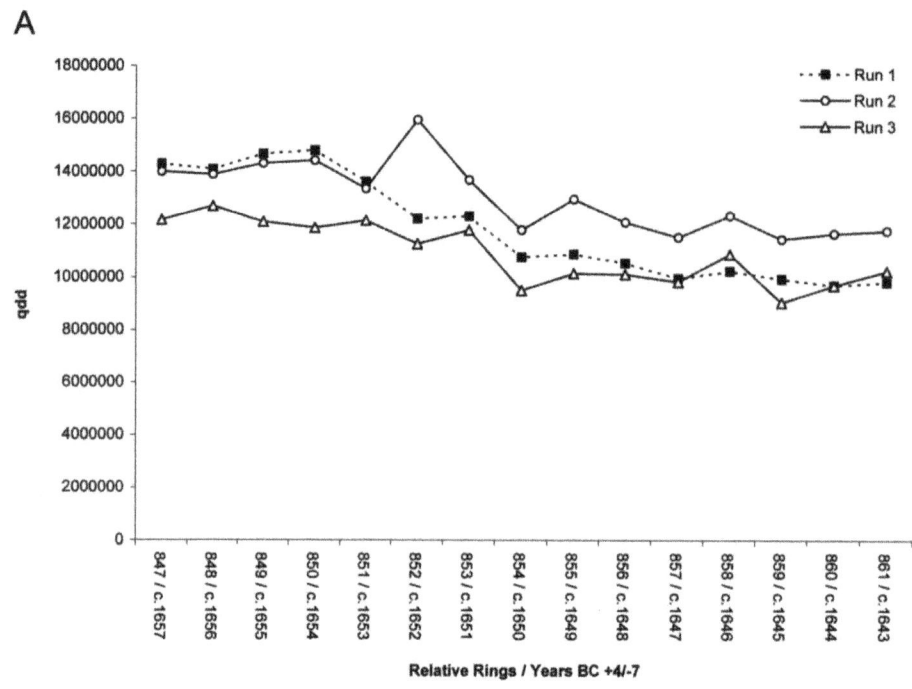
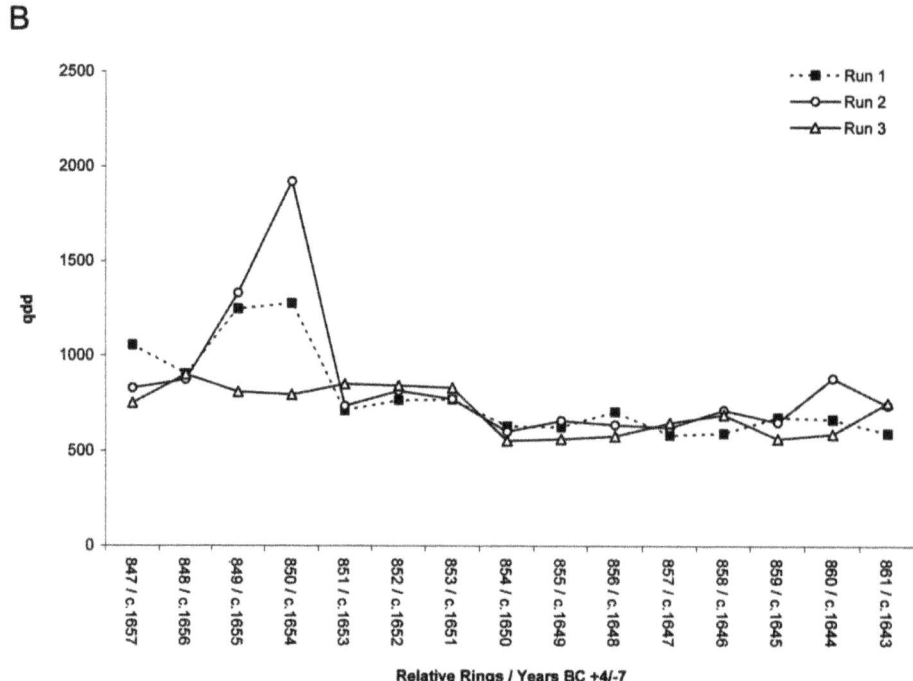

FIGURE 6.17: Example of poorly correlating concentration patterns for Ca and Mn in three replicate sequences from Por-Tu-26A. A is Ca, B is Mn

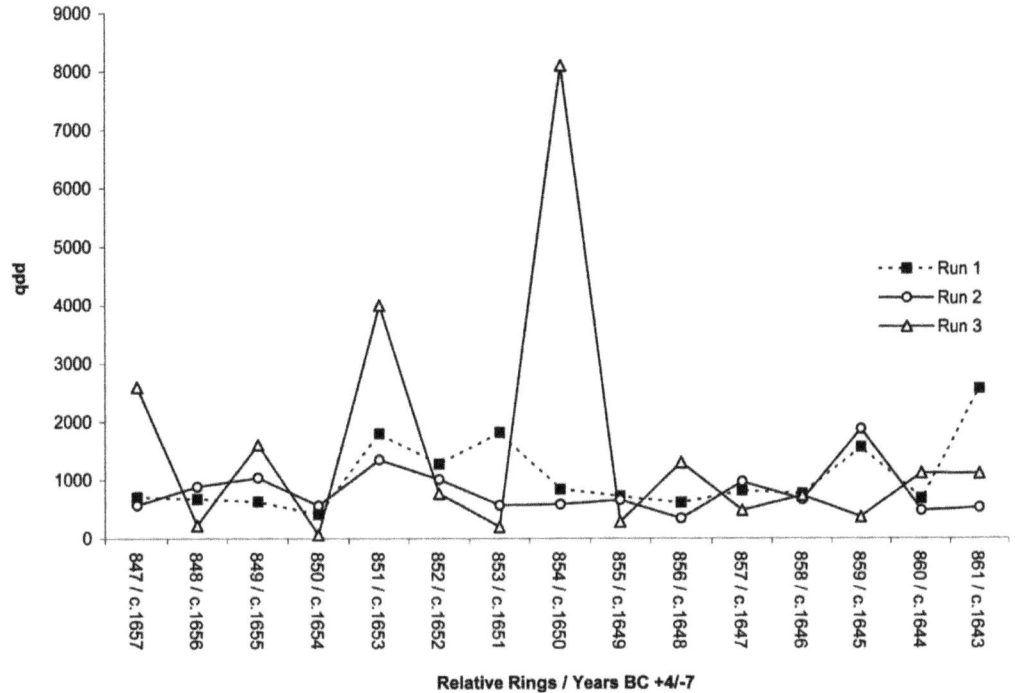

FIGURE 6.18: Example of how Zn failed to correlate for three replicate sequences from Por-Tu-26A

The other advantage to running the solutions via ICP-MS was that it was possible to expand the range of elements detected. In addition to the ICP-AES list, Li, V, Cr, Co, Ni, As, Rb, Sr, Y, Rh, Cd, Sb, Cs, La, Ce and Pb were detected. Of these Co, Ni, As, Rb, Sr, Y, and Rh displayed very similar concentration patterns to Mn and Ca etc. (see figures 6.6 and 6.8). Of further note are Li and Cr (figure 6.20), both of which show a fall in concentration after 854, and La and Ce which show a similar depletion proceeded by a rise (figure 6.21).

A separate sequence of samples for 26A were also prepared specifically for analysis by ICP-MS (according to the developed protocol for the analysis of wood discussed in section 3.4). These samples were four times the concentration of the ICP-AES samples, in order to try and detect a full suite of rare earth elements. However this meant that these samples were at the outer limit of the required levels for total dissolved solids in ICP-MS. The result of this was a reduction in the quality of the data produced as several correction steps were added to the data processing procedure to compensate for the steep drift experienced as the solids precipitated onto the sampling cone. The results displayed a distinct lack of correlation with those obtained via the analysis of the ICP-AES solutions. This could reflect the previously discussed problems (any further ICP-MS analysis of these samples would involve further dilution), or could simply represent sample heterogeneity.

By way of further comparison, a sub-section of the sampled sequence for 3A was analysed by LA-ICP-MS. Two complete and two partial radii were analysed over two days, seventeen different elements were detected. The results highlighted problems with the LA-ICP-MS method discussed in section 3.3.3, in that for the majority of elements instrumental detection wavered over the two days. This meant that the average size of the ratios to ^{13}C for the majority of elements were not replicable from day-to-day, however the overall trends of the data sets were shown to be similar. An example of this for Mn is provided in figure 6.22.

Data were processed in ratios format as the procedure for pellet calibration was still in the developmental stage at the time of analysis. For Ca, Mg and Sr both patterns and ratios were

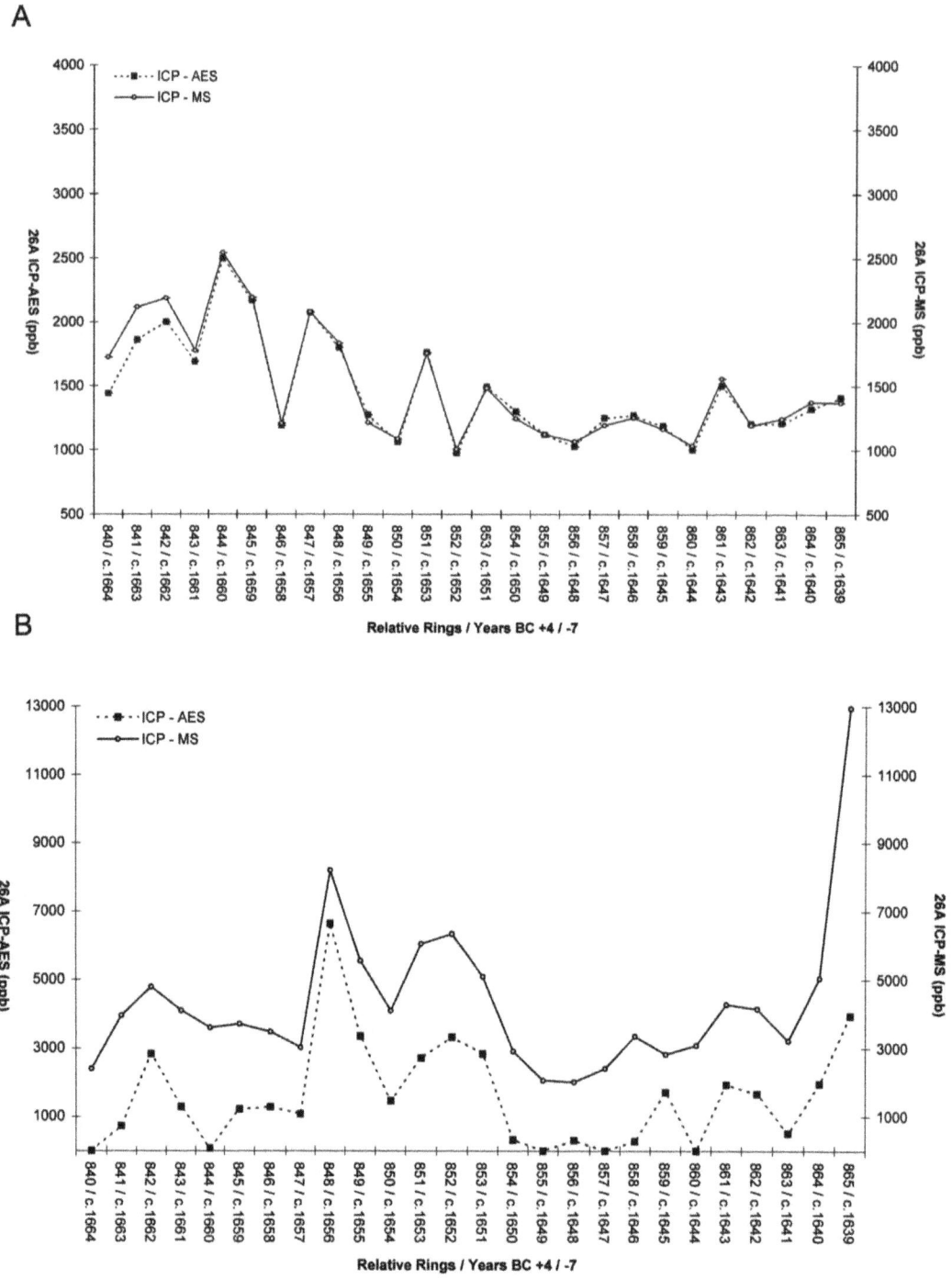

FIGURE 6.19: Examples of the degree of correlation between data generated for the same samples on two different instruments. Example A, Ba, shows good agreement, example B, Al shows the worst agreement. Error bars present but minimal

CHAPTER 6: CASE STUDY 2 - PORSUK, TURKEY

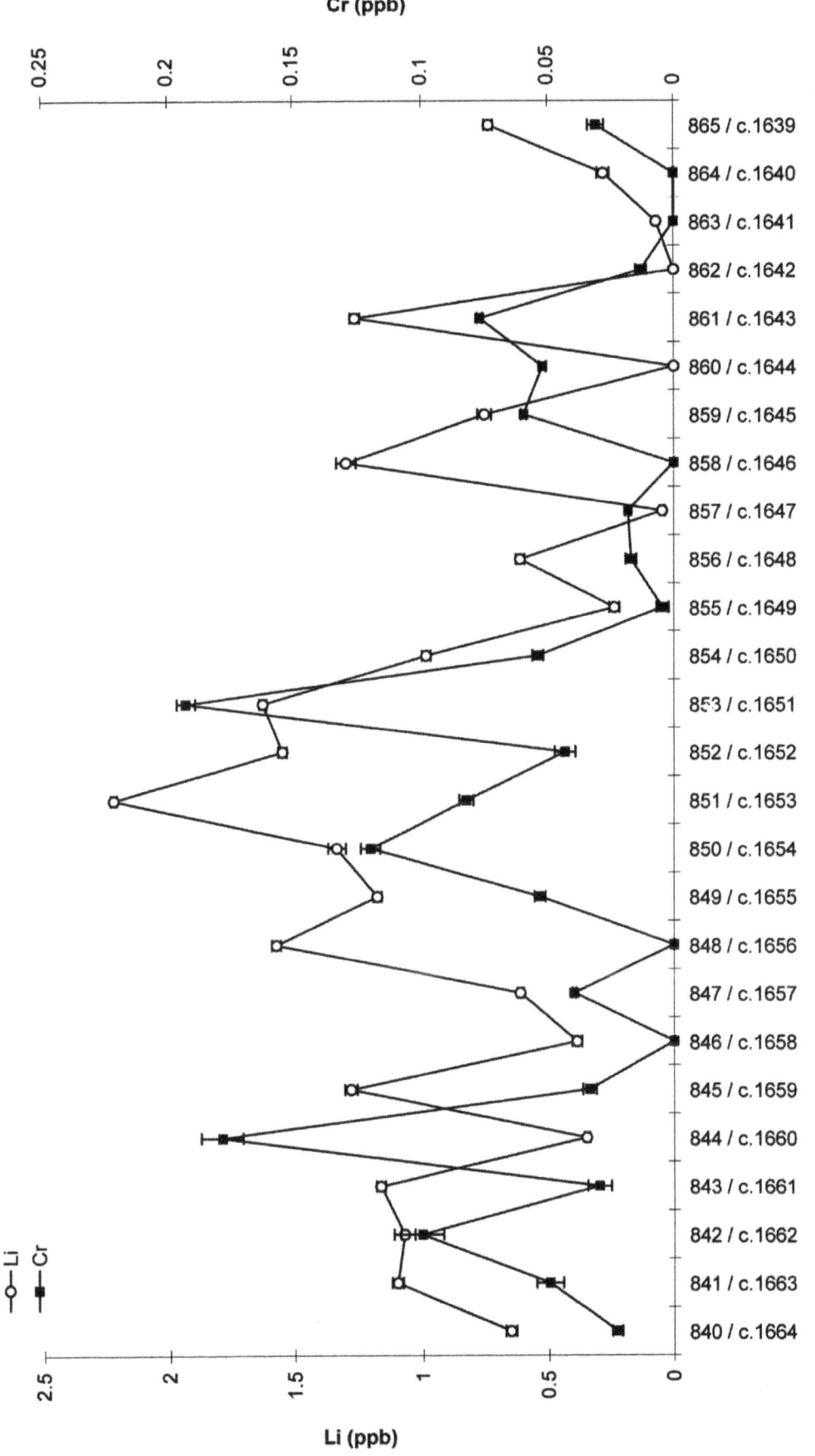

FIGURE 6.20: Li and Cr show a decrease in concentrations after relative ring 854

CHAPTER 6: CASE STUDY 2 - PORSUK, TURKEY

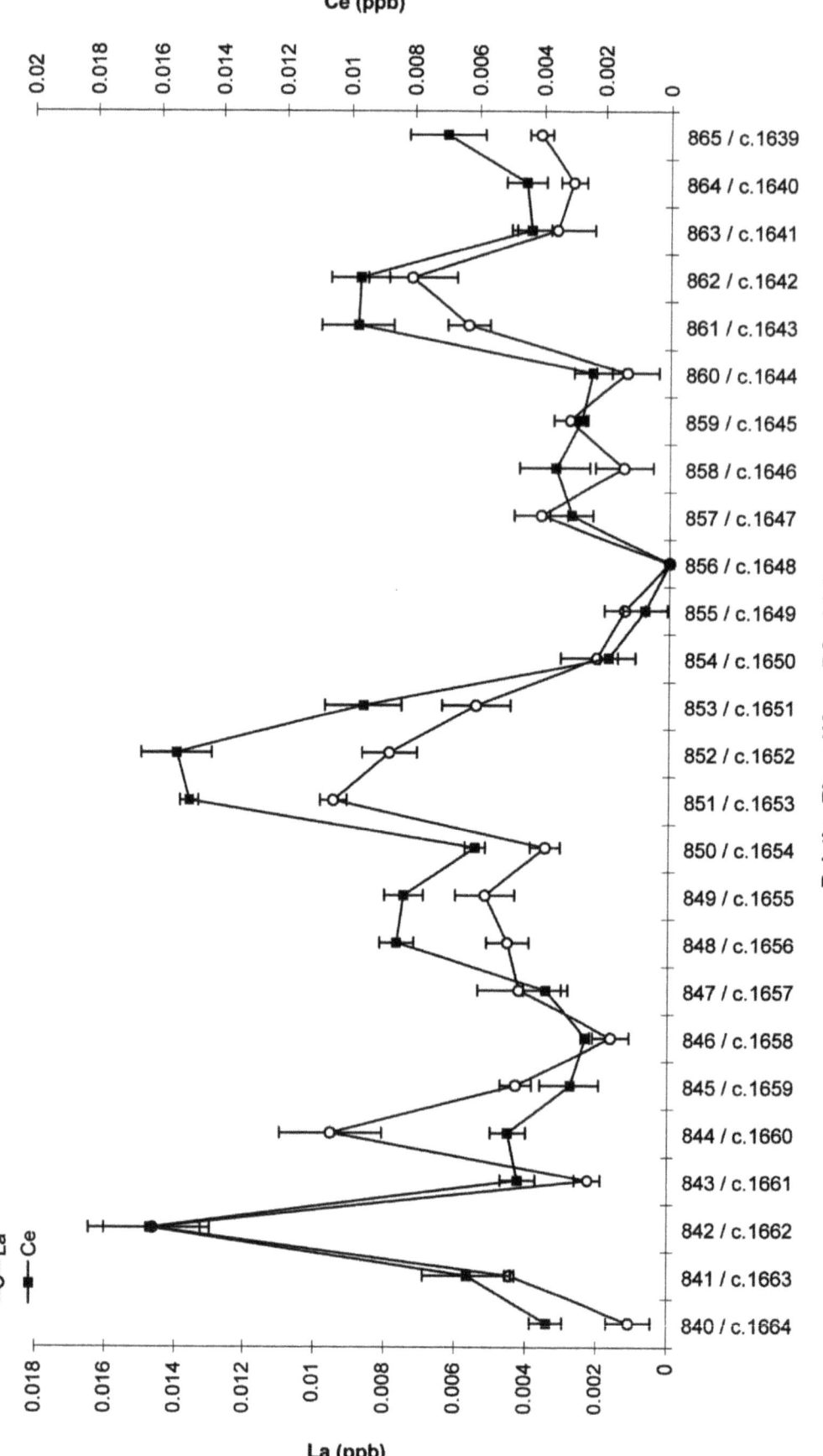

FIGURE 6.21: La and Ce show a decrease, preceded by a rise in concentrations around relative ring 854

CHAPTER 6: CASE STUDY 2 - PORSUK, TURKEY

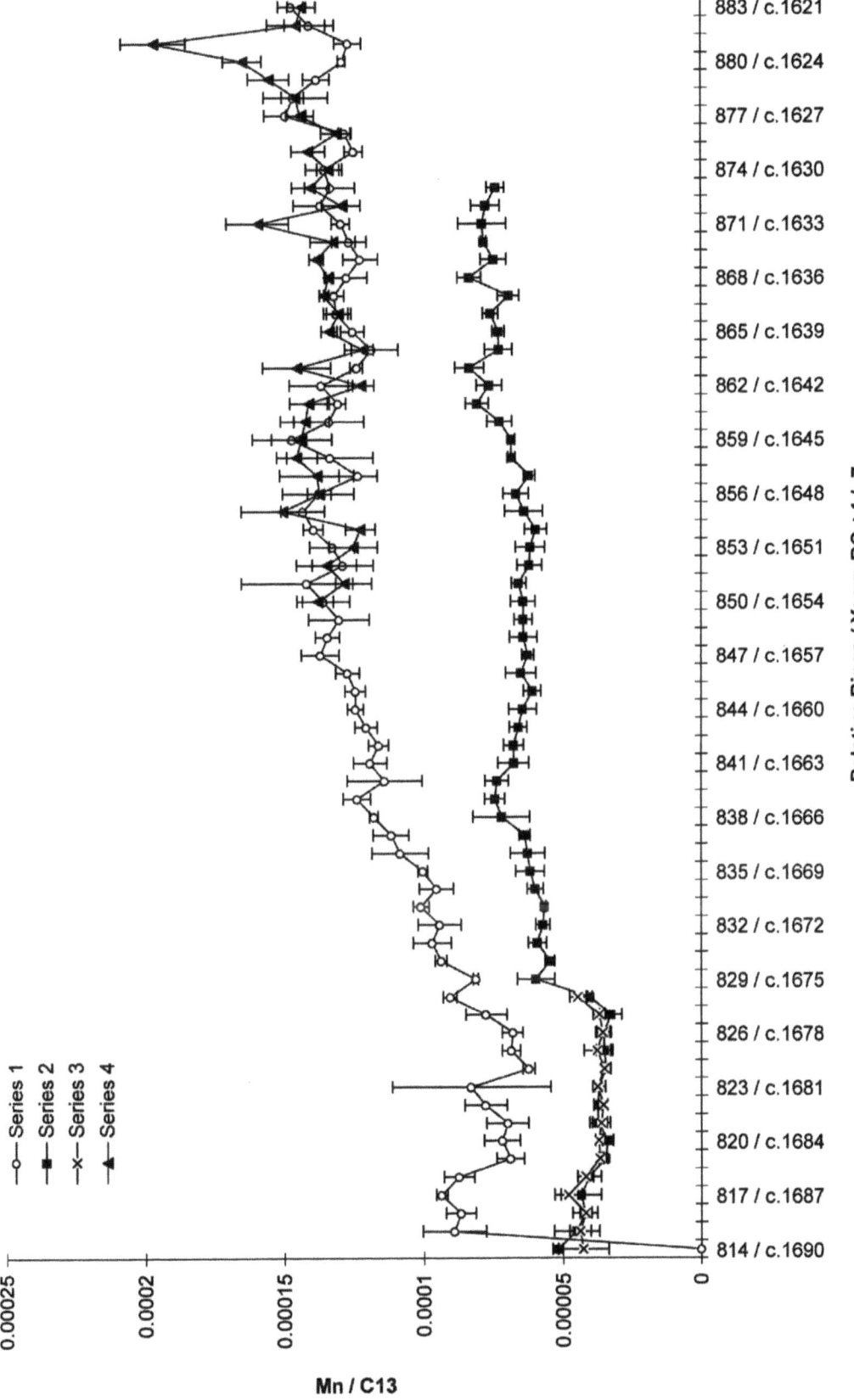

FIGURE 6.22: Mn ratio's to ^{13}C, over two days analysis. Whilst there is a difference between the size of the ratios for data sets collected on the two separate days the overall trend is very similar.

CHAPTER 6: CASE STUDY 2 - PORSUK, TURKEY

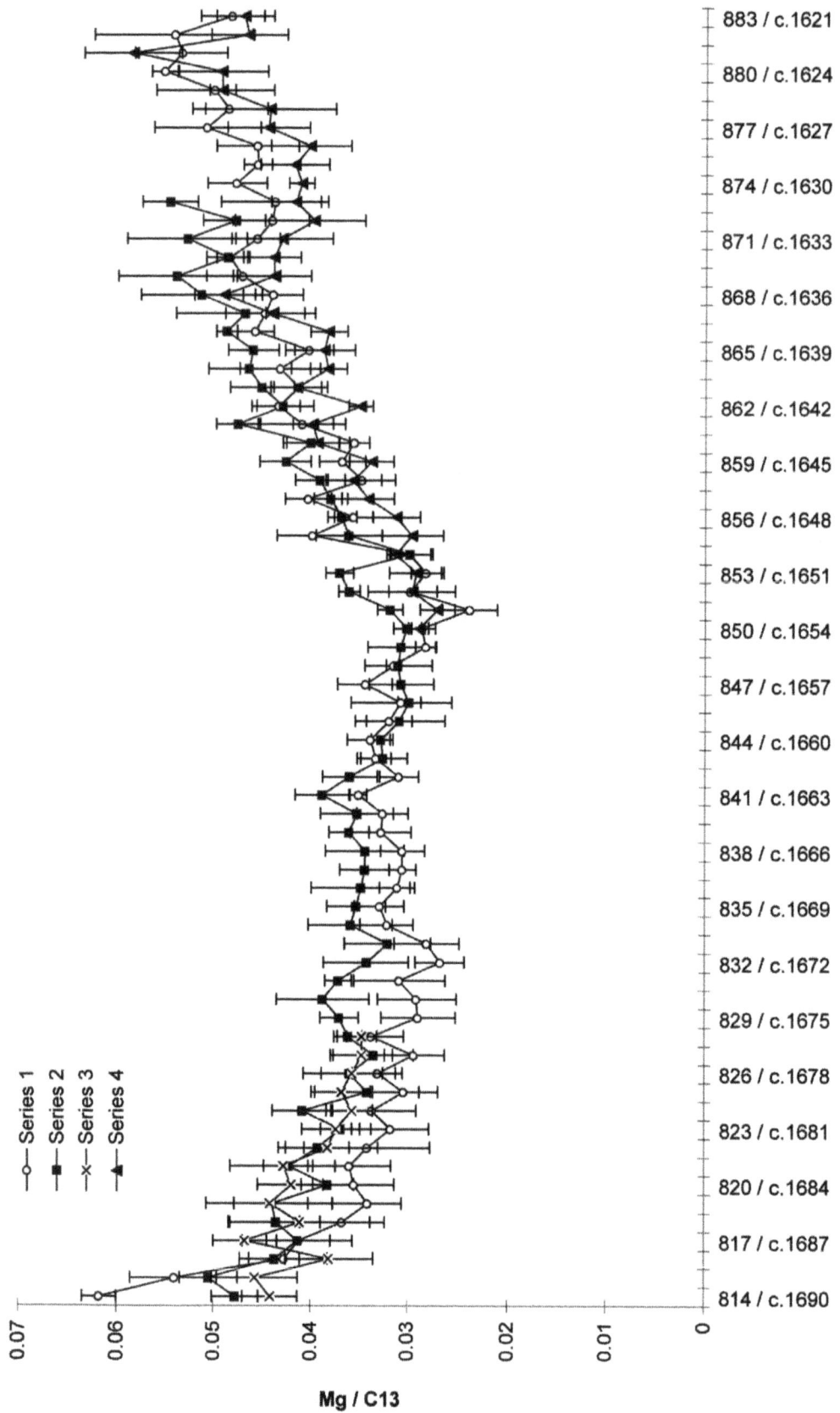

FIGURE 6.23: Mg ratio's to ^{13}C, over two days analysis. Ratios and trends correspond well

TABLE 6.2: Percentage average precision for each element from the same solutions via ICP-AES and ICP-MS. Where 'nd' the element was not determined via ICP-MS, Mg was detected but over saturated due to the unusually high concentrations

Element	ICP-AES (%)	ICP-MS (%)
Ag	14	28
Al	8	1
Ba	2	6
Ca	1	1
Cu	4	2
Fe	1	2
Mg	1	-
Mn	1	2
Na	3	nd
S	1	nd
Sr	2	1
Zn	5	9
Zr	3	0

replicable for both days of analysis. An example is given for Mg in figure 6.23

These elements also showed the best degree of agreement with data generated for the same years via ICP-AES. Examples for Ca and Sr are shown on figures 6.24 and 6.25. The agreement displayed for the majority of the elements for the same years via the two different techniques was good, with the three best correlations all showing a rise in concentrations from ring 854 via both techniques. The day-to-day agreement between the overall patterns of the elements via LA-ICP-MS was also good. This is somewhat unusual in the light of the high degree of heterogeneity displayed by single years from various modern wood samples. The apparently homogenous chemistry of these tree rings seems to confirm that the previously discussed ICP-MS solutions were simply too concentrated.

Up to this point the results seemed reasonably promising in terms of detecting some kind of change in chemical signature associated with ring 854. With a number of elements shown to be altering concentration around this year in each of the samples in a replicable way. The main problem with linking this pattern to a response to some environmental event within the tree rings is the concentrations themselves. The concentrations of certain elements are up to several hundred thousand ppm in some cases (e.g. Ca), this is several orders of magnitude higher than has previously been found in wood (see section 2.5.3 and table 6.3). In fact, they are much closer to those averages expected for soils which, whilst varying greatly, range from tens to thousands of ppm [76], e.g. Mn c. 200-5000 ppm, Zn c. 10-500 ppm or Cu c. 1-500 ppm. In order to explore this further, ten rings of modern day Juniper from a similar growth environment to that of the samples, were analysed by solution ICP-MS. Table 6.3 shows the concentrations found for detectable elements alongside the averages detected in the archaeological samples.

This comparison shows that the sample values are way in excess of those found in equivalent modern wood. They are however less than might be expected for a soil, possibly reflecting the fact that contamination was probably via permeation of ground water laden with a soluble fraction from the soil. The high quantity of Ca complies with what is known of the calcareous burial environment. The seeming homogeneity of the tree rings (identified previously) also supports the hypothesis that the elements detected have little to do with the wood itself.

6.5 Conclusions

Chemical analysis of individual tree rings from samples of archaeologically preserved wood covering one of the major volcanic eruptions of prehistory was carried out via three complimentary analytical techniques.

CHAPTER 6: CASE STUDY 2 - PORSUK, TURKEY

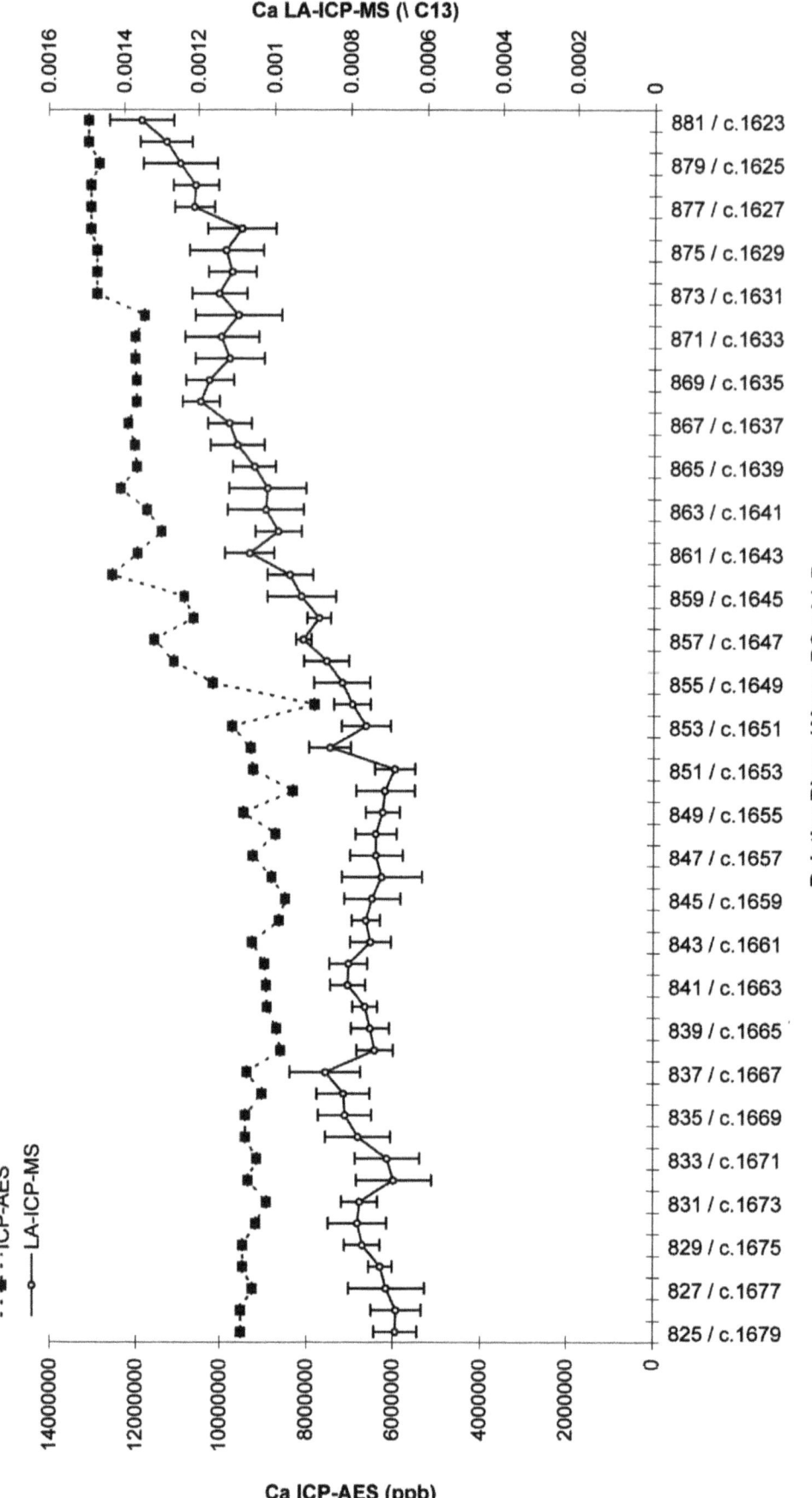

FIGURE 6.24: Ca elemental pattern from relative rings 825 - 881 via LA-ICP-MS and ICP-AES

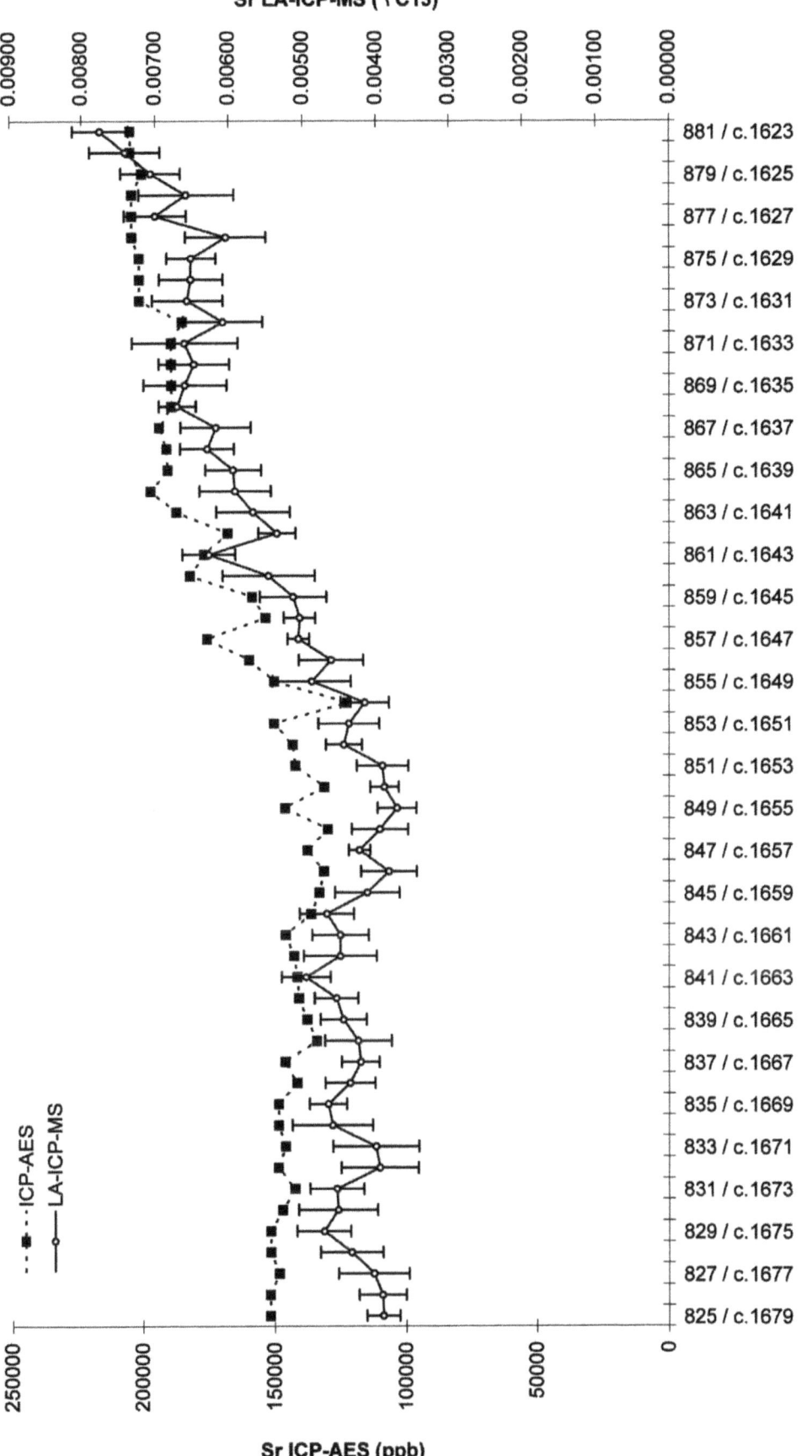

FIGURE 6.25: Sr elemental pattern from relative rings 825 - 881 via LA-ICP-MS and ICP-AES

TABLE 6.3: Average concentrations in the samples (ppb) versus average concentrations in a sample of modern juniper from a similar growth environment. Where 'nd', the element was not determined. (In the case of Mg the element was present but over saturated).

Element	Sample	Modern Juniper
Ag	29	BD
Al	1700	150
Ba	1700	30
Ca	11000000	34000
Cu	500	50
Fe	39000	35
Mg	1100000	-
Mn	490	22
Na	730000	nd
S	220000	nd
Sr	180000	108
Zn	1900	58
Zr	90	4

ICP-AES was used successfully in this context to detect S and other elements in wood. However, if the conclusion that the main derivation of the tree ring chemistry in this case is post-depositional, further analysis would be required to establish whether S and other elements from modern trees can be successfully detected by this method. For this particular sample, the developed methodology for the producing solutions for ICP-MS analysis would require further work in terms of addressing the total digestion of the sample (perhaps by the introduction of a microwave digestion step) and by ascertaining the optimum level of dilution. However, the lower concentration samples originally prepared for ICP-AES and then run via ICP-MS illustrate the fact that this technique can successfully detect a wider suite of elements than ICP-AES including some of the rare earth elements in the context of these particular samples. LA-ICP-MS was also found to work very successfully in regard to these samples. Sampling was largely non-destructive, preserving the ring width pattern and majority of the sub-sample for future analysis. Whilst a wide range of major elements were detected, the results were only fully replicable in terms of Ca, Mg and Sr on a day to day basis, however overall trends matched well for most elements. Corresponding general trends were detected for most overlapping elements for sample C-Tu-Por3A via both LA-ICP-MS and ICP-AES analysis, illustrating both the integrity of the detected elemental patterns and the reliability of the analytical procedures. The same points were illustrated for sample C-Tu-Por26A in running sub-samples of the same batch of solutions via both ICP-AES and ICP-MS. The three techniques were all used with varying degrees of success to both complement and cross check with one another. The various strengths of each could be tailored to a specific research question, for example, LA-ICP-MS can be used to ascertain the degree of heterogeneity in a sample and to gain an overview of the elemental chemistry of rare samples in a non-destructive way. The other two techniques can be used respectively to successfully detect different elements from samples which average out the chemistry of a single yearly increment.

As to the over-riding research objective, i.e. to ascertain if there is a chemical signature which may be linked to the Thera eruption, to be found in association with the Aegean dendrochronology's ring 854 anomaly, the results at this stage seen to indicate not. Beguiling though the apparent elemental changes around relative ring 854 are, the unusually high concentrations of the various elements and the relatively high degree of homogeneity displayed by the two samples clearly point to any true signature in the wood being drowned out by elemental uptake in the burial environment. The most obvious way to prove or disprove this hypothesis would be to obtain and analyse soil samples from the site in order to compare the elemental content and proportions of not only the soil itself, but of the more readily extractable elements which may have dominated the chemistry of water permeating the burial environment. If any elements were found in the wood but not in the soil samples it might be possible to begin to relate these to past environmental events, however given the apparent level of contamination of the overall sample this could never be substantiated.

6.6 Summary

This case study highlights some important points in the context of the application of dendrochemical techniques to dendrochronological samples. The findings underline the need to design a unique analytical protocol to address the specifics of any given sample. They emphasise the importance of developing a thorough understanding of the context of the sample and the nature of the environment of preservation, before commencing any interpretation. It was concluded that only tree ring samples from very specific preservation environments e.g. very cold and / or dry conditions could successfully be used in any future attempts to trace the volcanic eruptions of prehistory. In terms of actually attempting to trace the Thera eruption in these samples, the results were inconclusive. Whilst contamination in the burial environment seems the most likely explanation for all the patterns observed, perhaps further investigation of the true nature of the contamination would enable genuine patterns to be teased out from the background.

Chapter 7

Discussion

7.1 Introduction

This research project has investigated the potential of tree ring chemistry for the interpretation of volcanically derived signatures in absolutely dated annual growth increments. In doing so it has made substantial contributions to dendrochemical research in addition to furthering an archeological science approach to the issue of finding absolute dates for the major volcanic eruptions of pre-modern times. This has been achieved through three stages of research: (i), through a review of the discipline of dendrochemistry: (ii), through the development and review of both laser ablation and solution induction ICP-MS analytical techniques, and through the use of ICP-AES: (iii), through the use of the methodologies and analytical procedures developed to investigate a wide variety of wood species covering a variety of modern eruptions of known date, and an investigation of archaeological samples covering key volcanic events for which absolute dates are as yet unknown.

7.2 Dendrochemical review

Following an extensive search and in depth review of the relevant literature, the principles and potential problems inherent in all dendrochemical analysis were highlighted. From this it became clear that a thorough appreciation and understanding of all the likely complexities of dendrochemical research is essential before commencing any dendrochemical study. A need was identified for a means of collating and rapidly accessing the information gathered from the literature, specifically in terms of the behaviour of individual elements in various tree species. In order to meet this requirement, a database of reported observations was conceptualised, designed and continuously updated throughout the course of the project. The main purpose of the database was as an aid to the interpretation of subsequently generated data sets in the context of this project. However, it is also intended that the database be published in the form of an online resource available for anyone working in the field of dendrochemistry.

Whilst the review of the literature emphasised the complexity of the discipline, it also provided evidence to suggest that there were indeed some prospects for a dendrochemical resolution to the problem of finding an absolute date for past volcanic eruptions. Two main hypotheses were developed as to the type of chemical patterns one might logically or ideally hope to find as *volcanic signatures* in tree ring chemistry. Many studies have shown how increases in environmental acidity from various anthropogenic pollution sources have caused certain species of trees to absorb various elements more or less readily. It was hypothesised that acidity increases from volcanic eruptions could have exactly the same effect, and that various elements within the tree rings would respond with rises or falls in concentrations. It was also hypothesised that trees might directly absorb unusual volcanogenic elements, or combinations of elements. This might take the form of various concentrations of elements such as the rare earth elements, or ratios between isotopes of the same element (for example, Sr). This type of a signature was highlighted as the most desirable as only by defining a chemical finger print could a specific volcanic eruption be linked with a particular absolutely dated tree ring. Finally the literature was drawn upon to hone sampling and sample preparation strategies and to develop a new dendrochemical procedure to be applied to the further investigation of the main research question.

ysis via ICP-MS and ICP-AES. Solution ICP-MS was found to successfully detect qualitative concentrations for wide range of elements (including the rare earth elements) in sequences of individual tree rings. Problems identified with the technique included potential for the contamination of samples during the lengthy sample preparation process, and the fact that extremely narrow rings could not be analysed at annual resolution. ICP-AES was used successfully as a complimentary technique, specifically to detect S.

7.3 Methodological development

7.4 Results

New and important contributions were made to methodological design for the use of LA-ICP-MS, solution ICP-MS and ICP-AES in dendrochemical projects. The main areas of development were in terms of strategies for sampling, storing, cleaning, preparing and calibrating samples for LA-ICP-MS. Also for sampling, storing, cleaning, dissecting and digesting samples of absolutely dated tree rings for analysis by solution ICP-MS and ICP-AES. The use of LA-ICP-MS for the analysis of tree ring sequences was developed and reviewed. It was concluded that whilst the technique holds great potential for dendrochemical research in terms of the rapid multi-elemental analysis of long sequences of tree rings at sub-annual resolution, a solution induction methodology is likely to produce more workable data in the context of the main research question. There were two main reasons for this. First, that the best scenario for the project would be to detect a chemical figure print for a particular eruption in a tree ring. Unfortunately, the analyses consistently failed to detect a sufficiently wide range of elements, including important unusual and rare earth elements. Second, because the majority of wood samples tested appeared to be too heterogeneous[1] to be analysed by such a high resolution technique, it was considered that a more representative, average sample could be taken by solution analysis of a dissolved tree ring. A cheap, effective sample preparation methodology was designed (within the financial constraints of the project) by which sequences of rings from various tree species might be brought into solution for anal-

The results of the pilot studies and the two main case studies illustrated a number of outcomes. Overall, the results were used to test the success of the methodological development and to aid in the process of experimental design, however the focus in each case was to investigate the potential of the tree ring chemistry for detecting volcanic signatures. From the pilot studies, a number of data sets were produced which displayed possible responses by certain elements to volcanic eruptions. Pilot studies 1 and 2 (sections 4.2.1 and 4.3), hinted at the possibility of a response to the 1815 AD eruption of Tambora. In Case study 1, an apparent increase in the heterogeneity of certain tree rings around 1815 AD was linked to the formation of elemental hotspots due to large scale mobilisation of certain elements following a sudden increase in environmental acidity. Whilst the decision to use solution ICP-MS analysis as the primary analytical technique meant that this suspected heterogeneity could no longer be tested, it was anticipated that the average of the yearly chemistry provided by this technique would still reflect the raised elemental concentrations also associated with the hotspots. Whilst the results for Pilot study 2 were not so convincing as for Pilot study 1, they provided a useful comparison and a certain degree of agreement. Given that the results for the two studies were generated from totally different tree species grown in very different environments, the fact that both showed a possible response to Tambora is promising. Pilot study 3 (section 4.4) showed a possible association between a reduction in the presence

---

[1]It was concluded possible that the technique might be made to work more successfully on tree rings with more homogeneous chemistry such as *Acer sp.*

of certain elements (e.g. Mn, Sr and Rb) and a narrow ring event in the year 1350 thought to be due to the Kaharoa eruption. This may suggest that another proxy for detecting volcanic eruptions might be a reduction in the presence of certain elements due to a decrease in the functionality of the tree at this time.

Pilot studies 4 and 5 (sections 4.5 and 4.6) were of more use in terms of methodological development than in finding responses to volcanic eruptions. Pilot study 4 touched upon the subject of sample contamination in the burial environment, as issue which would prove central to the potential application of dendrochemical analysis to archaeological or palaeoenvironmentaly recovered samples, and which is considered in further detail in Case study 2.

The main aim of Pilot study 5 was to test ideas developed on future site selection, experimental design and sampling. It was a final attempt to use LA-ICP-MS to explore the main research objective. The data, as with the previous four studies, brought to light the key problems of sequence replication and the detection of an insufficient range of elements, especially the rare earth elements. From this data set it was finally concluded that solution ICP-MS would be a preferable mode of sample induction for most species of wood.

Pilot study 6 (section 4.7), put the methodology designed for the use of solution ICP-MS into practice for the first time. The results showed that a sequence of annual tree rings could be successfully analysed for the desired range of elements via solution ICP-MS. In addition, however, they raised important issues in respect to the detection of volcanic signatures. The analysed sequence covered no really major volcanic eruptions, yet the elemental patterns produced were exactly the sort one might link with a major eruption had one occurred in or around the years in which the pattern changed. It is true that minor eruptions occurred during the years covered by the sequence and many of these could be seen to offer plausible associations with the years in which these anomalies arose. However, a glance at Simkin and Siebert's chronology of the last 10,000 years of volcanism [142], shows that seldom a year goes by without tens of small scale volcanic eruptions occurring, therefore to simply claim linkage by apparent association would be totally unscientific. These results underlined the caution required in interpreting data sets covering the years of major volcanic eruptions. They highlighted the fact that only by identifying a specific elemental pattern or code in tree rings which can actually be matched to a similar pattern for a particular event can the matter be unequivocally resolved.

Case study 1 was the logical next step from pilot study 6. It put into practice the newly honed, completed methodology, not only for sample preparation and analysis via solution ICP-MS, but also for site selection and sampling. In view of the complex nature of dendrochemical analysis elucidated by the literature, it was decided that the best approach to completing a comprehensive dendrochemical study, was to obtain reliable data from the analysis of annual tree rings and then to consider all possible external influences which may have determined the various elemental patterns. The study, once again, illustrated that solution ICP-MS could successfully be used to detect the desired range of elements in tree rings. It also demonstrated that all aspects of sampling and sample preparation were working satisfactorily. However, no really conclusive links could be made with any of the major or local volcanic events covered within the time period sampled. The only possible signature found was in the 1991/1992 tree ring. A rise in a variety of elements was observed in two trees from the same site at this time. If this association could be connected to the Pinatubo eruption (1991) it would be the first step in beginning to look for evidence of past volcanism in tree rings. However, the lack of signal for the more major Tambora eruption, called even this association into question: why should one event show up and not the other? It was concluded that much of what was observed in the sequences was a type of *background noise* from a relatively favorable growth environment. Although the site was selected for its thin soils and relatively low background pollution, these results emphasised that even more care must be taken over site selection, in order to maximise the potential for signature detection. In particular it was concluded that the ideal site type would not only have the thin, nutrient poor soils previously specified, but would also be dominated by a more arid climate regime. Overall, the data presented underlined the complexity of the heterogeneous chemistry of the annual growth rings of *Pinus sylvestris*. However it also highlighted the fact that there is a lot of potential information to be gained from tree ring chemistry, and what remains is to improve our understanding of the processes of tree physiology as an aid to interpretation.

The second case study was completed with a more specific aim. This was to investigate how well the developed analytical methodology could be applied to an archaeological sample. The purpose for exploring this, was that if a volcanic signature can one day be positively proven in tree rings, the next step would be to attempt to apply this science to date key volcanic eruptions of unknown date. To do this would require the analysis of samples of dendrochronologically dated wood of substantial age. The sample in question had been preserved in a relatively dry environment, unlike the bog oak considered in Chapter 4, (section 4.5) so it seemed possible the original chemistry of the wood may have remained intact. However, the results showed that this sample had also been contaminated in the burial environment.

Conducting dendrochemical studies with relatively modern wood obtained from dendrochronologies, also comes with a range of problems. Instead of having all the site specific information required for a comprehensive dendrochemical study, often very little background information accompanies the samples. This is because sampling was initially conducted for the purpose of chronology building rather than for dendrochemical work. Likewise it cannot be ensured that samples were taken from the best possible height or aspect within the tree, and there is no control over the type of contamination which may have occurred between sampling and analysis. The further necessity of working with extremely old wood samples is fraught with a further array of potential problems. Such samples are often rare and there may not be sufficient to make several replicate sequences, and critically in order to have survived to the present day they must have undergone some form of preservation.

The findings from this case study emphasised the complexity involved with this type of study. They underlined the importance of developing a thorough understanding of the history of a sample, and the nature of the environment in which it was preserved, before commencing any analysis and interpretation. In view of this the eventual prospects for pinning down an absolute date for key volcanic events such as the Thera eruption may not seem particularly hopeful. However, given the right combination of circumstances (for example where samples were preserved in very cold and / or very dry conditions) and the design of a unique analytical protocol to get the best from any given sample, it may still be possible. The hints at chemical associations with certain eruptions, derived from the various analyses carried out over the duration of this project, were sufficiently encouraging to suggest that it may, given a logical approach and the right set of circumstances, be possible for specific events to be identified in specific pieces of wood.

7.5 Conclusions

The primary research objective was explored through the three stages of research introduced in section 1.3 of the introduction and discussed previously. The main outcomes of each stage of research and the conclusions reached will now be summarised. The main conclusions are highlighted in italics.

1. Dendrochemical review

    (a) A review was made of all available literature pertaining to aspects of dendrochemical research.

    (b) A need was identified for a means of collating information from the literature into a readily accessible source.

    (c) A database was designed and maintained to meet this need.

    (d) Evidence was found to show that annual concentrations of specific elements in tree rings can, in certain scenarios, directly reflect changes in external environmental chemistry.

    (e) Evidence was found to support hypotheses for how particular volcanic signatures might be observed in tree rings.

    (f) The literature was used as an aid to honing the methodological development.

(g) *The chemistry of various species of tree was found to be highly complex, variable, and effected by many physiological and environmental factors.*

(h) *From the existing literature it was concluded to be possible that a volcanic signature of some sort may be found in certain tree rings.*

2. Method development

(a) An optimised methodology was designed and applied for site selection, tree selection and sampling.

(b) An effective new methodology was designed and applied for sub-sampling, storing, cleaning and preparing samples for LA-ICP-MS.

(c) Various attempts were made at new calibration methodologies for LA-ICP-MS.

(d) An effective new methodology was designed and applied for sub-sampling, storing, cleaning, dissecting and digesting samples for analysis by solution ICP-MS and ICP-AES.

(e) Both techniques (along with ICP-AES) were used to successfully analyse sequences of tree rings, within the limitations of each technique.

(f) LA-ICP-MS was found to have great potential use for the rapid, high resolution, multi-elemental analysis of tree ring sequences.

(g) *The current level of calibration and a failure to replicate sequences due to the heterogeneity of the tree rings, led to the rejection of LA-ICP-MS as a suitable technique with which to pursue the main research objective. However, future use of a higher resolution and more sensitive LA-ICP-MS could allow mapping of hotspots and anatomical associations with higher concentrations. In this way the technique holds further potential in terms of investigating the true nature of the heterogeneity.*

(h) *Solution induction ICP-MS was selected as the preferable analytical technique with which to continue analysis.*

3. New analyses

(a) 2300 laser ablation samples, and 500 solution ICP-MS and ICP-AES samples were run.

(b) The majority of the samples were concerned with the development of the methodology, however over 800 samples were part of pilot and case studies to investigate the chemistry of tree rings in years when eruptions were known to have occurred.

(c) Aims of the case studies were not only to explore tree ring chemistry but also to assess the possible complications of applying dendrochemical techniques to ancient wood samples.

(d) *Some promising results were obtained which appear to relate certain chemical anomalies in tree rings to certain volcanic eruptions.*

(e) *Given the time constraints set upon this project, it was not possible to obtain a significant degree of replication for any of these sequences.*

(f) *The results highlight the complexity of dendrochemical studies and the care with which data must be interpreted, also the potential offered by tree ring chemistry if it could be better understood.*

7.6 Concluding remarks

The multi-elemental analysis of individual tree rings has been thoroughly explored, with a view to investigating the potential application of dendrochemical techniques for elucidating absolute dates for pre-modern volcanic eruptions. It is concluded that the living system which makes up a tree is complex and poorly understood. The chemistry of the individual tree rings within this system is also, as a result, complex and poorly understood. However, given the right tree, the right growth environment and the right environmental conditions, dendrochemical studies have shown that annual concentrations of specific elements in tree rings, can directly reflect changes in external environmental chemistry. Moreover, there is tentative evidence both from the existing literature and from the findings presented in this study, that changes in environmental chemistry related to volcanic eruptions may be similarly detectable. However, what has also been highlighted by this study is just how difficult this association will be to define and quantify. Whether or not a particular eruption shows up in the chemistry of a tree ring relies upon a massive number of variables. Add to this the further complications of obtaining ancient samples from dendrochronological sequences and the picture becomes even more complicated. Further research should focus on very specifically designed studies which take into full consideration all aspects of the history of the sample in question. These studies should be on suitable samples of wood, grown in marginal, dry environments and preserved by very cold or dry conditions. Above all, the search for a chemical finger print should be pursued as the only undisputable means of connecting a certain eruption with a particular tree ring, and this can only be truly substantiated by thorough replication. Given time, the right samples and the continuous improvement of analytical procedures such as solution ICP-MS, it may be possible to date certain volcanic eruptions via the chemistry of individual tree rings, but it seems unlikely that this approach will ever, once and for all, be irrefutably resolved.

Bibliography

[1] Adams, J. B., Mann, M. E. and Ammann, C. M. (2003). "Proxy evidence for an El Nino-like response to volcanic forcing." *Nature*, **vol. 426**:274–278.

[2] Alteyrac, J., Augagneur, S., Médina, B., Vivas, N. and Glories, Y. (1995). "Method of determination of mineral concentration in oak wood by laser ablation-ICP-MS." *Analusis*, **vol. 23**:523–526.

[3] Amato, I. (1988). "Tapping tree rings for the environmental tales they tell." *Analytical Chemistry*, **vol. 60**:1103A–1107A.

[4] Aoki, T., Katayama, Y., Kagawa, A., Koh, S. and Yoshida, K. (1998). "Measurement of trace elements in tree rings using the PIXE method." *Nuclear Instruments and Methods in Physics Research B*, **vol. 136**:919–922.

[5] Baes, C. F., III and McLaughlin, S. B. (1984). "Trace elements in tree rings: evidence of recent and historical air pollution." *Science*, **vol. 224**:494–497.

[6] Bailey, J. H. E. and Reeve, D. W. (1993). "Determination of the spatial distribution of trace elements in black spruce, *Picea mariana* Mill., by imaging microprobe secondary ion mass spectrometry." *International Symposium on Wood and Pulping Chemistry*, **vol. 2**:848–857.

[7] Bailey, J. H. E. and Reeve, D. W. (1996). "Determination of the spatial distribution of trace elements in jack pine, *Pinus banksiana* Lamb., by imaging microprobe secondary ion mass spectrometry." *Journal of Pulp and Paper Science*, **vol. 22**:274–278.

[8] Baillie, M. G. L. (1995). A Slice Through Time. B.T. Batsford Ltd, London.

[9] Baillie, M. G. L. (1998). "Bronze age myths expose archaeological shortcomings? a reply to Buckland *et al.* 1997." *Antiquity*, **vol. 72**:425–432.

[10] Baillie, M. G. L. (1999). Exodus to Arthur: catastrophic encounters with comets. B.T.Batsford, London.

[11] Baillie, M. G. L. and Munro, M. A. R. (1988). "Irish tree rings, Santorini and volcanic dust veils." *Nature*, **vol. 332**:344–346.

[12] Barci-Funel, G., Dalmasso, J., Barci, V. L. and Ardisson, G. (1995). "Study of the transfer of radionuclides in trees at a forest site." *The Science of the Total Environment*, **vol. 173**:369–373.

[13] Bekki, S. and Pyle, J. A. (1994). "A two-dimensional modeling study of the volcanic eruption of Mount Pinatubo." *Journal of Geophysical Research*, **vol. 99**:18861–18869.

[14] Berglund, A., Brelid, H., Rindby, A. and Engstrom, P. (1999). "Spatial distribution of metal ions in spruce wood by synchrotron radiation microbeam X-ray flourescence analysis." *Holzforschung*, **vol. 53**:474–480.

[15] Berninghausen, W. H. (1964). "A checklist of Icelandic volcanic activity." *Bulletin of the Seismological Society of America*, **vol. 54**:443–450.

[16] Bogaard, P. V., Hall, C. M., Schmincke, H. U. and York, D. (1989). "Precise single grain 40Ar/39Ar dating of a cold to warm climatic transition in Central Europe." *Nature*, **vol. 342**:523–525.

[17] Bondietti, E. A., Baes, C. F., III and McLaughlin, S. B. (1989). "Radial trends in cation ratios in tree rings as indicators of the impact of atmospheric deposition on forests." *Canadian Journal of Forest Research*, **vol. 19**:586–594.

[18] Bondietti, E. A., Momoshima, N., Shortle, W. C. and Smith, K. T. (1990). "A historical perspective on divalent cation trends in red spruce stemwood and the hypothetical relationship to acidic deposition." *Canadian Journal of Forest Research*, **vol. 20**:1850–1858.

[19] Brabander, D. J., Keon, N., Stanley, R. H. R. and Hemond, H. F. (1999). "Intra-ring variability of Cr, As, Cd, and Pb in red oak revealed by secondary ion mass spectrometry: Implications for environmental biomonitoring." *Proceedings of the National Academy of Science USA*, **vol. 96**:14635–14640.

[20] Briffa, K. R., Jones, P. D., Schweingruber, F. H. and Osborn, T. J. (1998). "Influences of volcanic eruptions on northern hemisphere summer temperature over the past 600 years." *Nature*, **vol. 393**:450–457.

[21] Briffa, K. R., Osborn, T. J., Schweingruber, F. H., Jones, P. D., Shiyatov, S. G. and Vaganov, E. A. (2002). "Tree-ring width and density data around the Northern Hemisphere: Part 1, local and regional climate signals." *The Holocene*, **vol. 12**:737–757.

[22] Briffa, K. R., Osborn, T. J., Schweingruber, F. H., Jones, P. D., Shiyatov, S. G. and Vaganov, E. A. (2002). "Tree-ring width and density data around the Northern Hemisphere: Part 2, spatio-temporal variability and associated climate patterns." *The Holocene*, **vol. 12**:759–789.

[23] Buckland, P. C., Dugmore, A. J. and Edwards, K. J. (1997). "Bronze age myths? volcanic activity and human response in the Mediterranean and North Atlantic regions." *Antiquity*, **vol. 71**:581–593.

[24] Cadogan, C. (1988). "Dating of the Santorini eruption (Cadogan replies)." *Nature*, **vol. 332**:401–402.

[25] Candelone, J. P. and Hong, S. (1995). "Post Industrial Revolution changes in large-scale atmospheric pollution of the northern hemisphere by heavy metals as documented in central Greenland snow and ice." *Journal of Geophysical Research*, **vol. 100**:16605–16616.

[26] Chenoweth, M. (2001). "Two major volcanic cooling episodes derived from global marine air temperature, AD 1807-1827." *Geophysical Research letters*, **vol. 28**:2963–2966.

[27] Chisholm, W., Rosman, K. J. R., Candelone, J., Boutron, C. F. and Bolshov, M. A. (1997). "Measurement of bismuth at pg g-1 concentrations in snow and ice samples by thermal ionisation mass spectrometry." *Analytica Chimica Acta*, **vol. 347**:351–358.

[28] Chun, L. and Hui-yi, H. (1992). "Tree-ring element analysis of Korean pine (*Pinus koraiensis* Siel. Et Zuc.) and Mongolian oak (*Quercus mongolica* Fisch. Ex Turcz. from Changbai Mountain, north east China." *Trees*, **vol. 6**:103–108.

[29] Clausen, H. B., Hammer, C. U., Hvidberg, C. S., Dahl-Jensen, D., Steffensen, J. P., Kipfstuhl, J. and Legrand, M. (1997). "A comparison of the volcanic records over the past 4000 years from the Greenland Ice Core Project and Dye 3 Greenland Ice Cores." *Journal of Geophysical Research*, **vol. 102**:26707–26723.

[30] Coffey, M. T. (1996). "Observations of the impact of volcanic activity on stratospheric chemistry." *Journal of Geophysical Research*, **vol. 101**:D3 6767–6780.

[31] Conkey, L. E. (1986). "Red spruce tree-ring widths and densities in Eastern North America as indicators of past climate." *Quaternary Research*, **vol. 26**:232–243.

[32] Cutter, B. E. and Guyette, R. P. (1993). "Anatomical, chemical, and ecological factors affecting tree species choice in dendrochemistry studies." *Journal of Environmental Quality*, **vol. 22**:611–619.

[33] D'Arrigo, R. D. and Jacoby, G. C. (1999). "Northern North American tree-ring evidence for regional temperature changes after major volcanic events." *Climatic Change*, **vol. 41**:1–15.

[34] Dean, J. S., Meko, D. M. and Swetnam, T. W. (1994). Tree rings, environment and humanity. Department of Geosciences, Tucson, Arizona.

[35] Delmas, R. J. and Legrand, M. (1998). "Trends recorded in Greenland in relation with Northern Hemisphere anthropogenic pollution." *IGAC Activities Newsletter*, **vol. 14**:1–6.

[36] DeWalle, D. R., Swistock, B. R., Sayre, R. G. and Sharpe, W. E. (1991). "Spatial variations of sapwood chemistry with soil acidity in Appalachian forests." *Journal of Environmental Quality*, **vol. 20**:486–491.

[37] Donnelly, J. R., Shane, J. B. and Schaberg, P. G. (1990). "Lead mobility within the xylem of red spruce seedlings: implications for the development of pollution histories." *Journal of Environmental Quality*, **vol. 19**:268–271.

[38] Duff, P. M. D. (1994). Holme's Principles of Physical Geology, Forth Edition. Chapman & Hall, London.

[39] Dugmore, A. (1989). "Icelandic volcanic ash in Scotland." *Scottish Geographical Magazine*, **vol. 105**:168–172.

[40] Ek, A. S., Lofgren, S., Bergholm, J. and Qvarfort, U. (2001). "Environmental effects of one thousand years of copper production at Falun, central Sweden." *Ambio*, **vol. 30**:96–103.

[41] Eklund, M. (1995). "Cadmium and lead deposition around a Swedish battery plant as recorded in oak tree rings." *Journal of Environmental Quality*, **vol. 24**:126–131.

[42] Falken-Gerup, U., Linnermark, N. and Tyler, G. (1987). "Changes in acidity and cation pools of south Swedish soils between 1949 and 1985." *Chemosphere*, **vol. 16**:2239–2248.

[43] Ferguson, C. W. (1969). "A 7104 year annual tree ring chronology for bristle cone pine, *Pinus aristata*, from the White Mountains, California." *Tree Ring Bulletin*, **vol. 29**:2–29.

[44] Filion, L., Payette, S., Gauthier, L. and Boutin, Y. (1986). "Light rings in subarctic conifers as a dendrochronological tool." *Quaternary Research*, **vol. 26**:277–279.

[45] Freer-Smith, P. H. and Read, D. B. (1995). "The relationship between crown condition and soil solution chemistry in oak and sitka spruce in England and Wales." *Forest Ecology and Management*, **vol. 79**:185–196.

[46] Frelich, L. E., Bockheim, J. G. and Leide, J. E. (1988). "Historical trends in tree-ring growth and chemistry across an air-quality gradient in Wisconsin." *Canadian Journal of Forest Research*, **vol. 19**:112–121.

[47] Fritts, H. C. (1976). Tree Rings and Climate. Academic Press, London.

[48] Gervais, B. R. and MacDonald, M. (2001). "Tree-ring and summer-temperature response to volcanic aerosol forcing at the Northern tree-line, Kola Peninsula, Russia." *The Holocene*, **vol. 11**:499–505.

[49] Giuffre, G. P. and Litman, R. (1979). "Comparison of elemental uptake by pine trees in varied environments." *Journal of Environmental Science Health*, **vol. 5**:365–375.

[50] Glass, G. A., Hasenstein, K. H. and Chang, H. T. (1993). "Determination of trace element concentration variations in tree rings using PIXE." *Nuclear Instruments and Methods in Physics Research B*, **vol. 79**:393–396.

[51] Grattan, J. (1995). "The distal impact of Icelandic volcanic gases and aerosols in Europe: a review of the 1783 Laki Fissure eruption and environmental vulnerability in the late 20th century." *Engineering Geology Special Publication*, **vol. 15**:97–104.

[52] Graumlich, L. J. and Brubaker, L. B. (1986). "Reconstruction of annual temperature (1590-1979) for Longmire, Washington, derived from tree rings." *Quaternary Research*, **vol. 25**:223–234.

[53] Grudd, H., Briffa, K. R., Gunnarson, B. E. and Linderholm, H. W. (2000). "Swedish tree rings provide new evidence in support of a major, widespread environmental disruption in 1628BC." *Geophysical Research Letters*, **vol. 27**:2957–2960.

[54] Guyette, R. P., Cutter, B. E. and Henderson, G. S. (1991). "Long-term correlations between mining activity and levels of lead and cadmium in tree-rings of eastern red-cedar." *Journal of Environmental Quality*, **vol. 20**:146–149.

[55] Guyette, R. P., Cutter, E. C. and Henderson, G. S. (1989). "Long-term relationships between Molybdenum and Sulfur concentrations in redcedar tree rings." *Journal of Environmental Quality*, **vol. 18**:385–389.

[56] Haas, G. and Muller, A. (1995). "Radioecological investigations on tree rings of spruce." *The Science of the Total Environment*, **vol. 173**:393–397.

[57] Hagemeyer, J., Lulfsmann, A., Perk, M. and Breckle, S. W. (1992). "Are there seasonal variations of trace element concentrations (Cd, Pb, Zn) in wood of *Fagus* trees in Germany?" *Vegetatio*, **vol. 101**:55–63.

[58] Hall, G. S., Yamaguchi, D. K. and Rettberg, T. M. (1990). "Multielemental analysis of tree rings by inductively coupled plasma mass spectrometry." *Journal of Radioanalytical Nuclear Chemistry Letters*, **vol. 146**:255–265.

[59] Hall, V. A. and Pilcher, J. R. (2002). "Late-Quaternary Icelandic tephras in Ireland and Great Britain: detection, characterization and usefulness." *The Holocene*, **vol. 12**:223–230.

[60] Hammer, C. U., Clausen, H. B. and Dansgaard, W. (1980). "Greenland ice sheet evidence of post-glacial volcanism and its climatic impact." *Nature*, **vol. 288**:230–235.

[61] Hammer, C. U., Clausen, H. B., Fredrich, W. L. and Tauber, H. (1988). "Dating of the Santorini eruption (Hammer *et al* reply)." *Nature*, **vol. 332**:401–402.

[62] Hammer, C. U., Clausen, H. B., Friedrich, W. L. and Tauber, H. (1987). "The Minoan eruption of Santorini in Greece dated to 1645BC?" *Nature*, **vol. 328**:517–519.

[63] Häsänen, E. and Huttunen, S. (1989). "Acid deposition and the element composition of pine tree rings." *Chemosphere*, **vol. 18**:1913–1920.

[64] Hinterstoisser, B., Hofinger, A., Hofer, S., Ulreich, M. and Wimmer (1996). "NMR, FTIR and HPLC techniques as analytical tools in tree-ring investigations." In: "Tree rings, environment and humanity," , editors Dean, J. S., Meko, D. M. and Swetnam, T. W. Radiocarbon, Department of Geosciences, Tucson, Arizona.

[65] Hoffmann, E., Ludke, C., Scholze, H. and Stephanowitz, H. (1994). "Analytical investigations of tree rings by laser ablation ICP-MS." *Fresenius Journal of Analytical Chemistry*, **vol. 350**:253–259.

[66] Hoffmann, E., Stephanowitz, H. and Skole, J. (1996). "Investigations of the migration of elements in tree rings by laser-ICP-MS." *Fresenius Journal of Analytical Chemistry*, **vol. 355**:690–693.

[67] Hora, B. (1981). The Oxford encyclopedia of trees of the world. Oxford University Press, Oxford.

[68] Hughes, M. K. (1988). "Ice-layer dating of eruption at Santorini." *Nature*, **vol. 335**:211–212.

[69] Hutchinson, T. C., Watmough, S. A., Sager, E. P. S. and Karagatzides, J. D. (1998). "Effects of excess nitrogen deposition and soil acidification on sugar maple (*Acer saccharum*) in Ontario, Canada: an experimental study." *Canadian Journal of Forest Research*, **vol. 28**:299–310.

[70] Huttunen, S., Karhu, M. and Torvela, H. (1983). "Deposition of suphur compounds on forests in southern Finland." *Aquilo se Botanica*, **vol. 19**:270–274.

[71] Jarvis, K. E., Gray, A. L. and Houk, R. S. (2003). Handbook of Inductively Coupled Plasma Mass Spectrometry. Viridian Publishing, London.

[72] Jensen, A. (1997). "Historical deposition rates of Cd, Cu, Pb, and Zn in Norway and Sweden estimated by 210Pb dating and measurement of trace elements in cores of peat bogs." *Water, Air and Soil pollution*, **vol. 95**:205–220.

[73] Jones, P. D., Osborn, T. J. and Briffa, K. R. (2001). "The evolution of climate over the last millennium." *Science*, **vol. 292**:662–667.

[74] Jonsson, A., Eklund, M. and Hakansson, K. (1997). "Heavy metals of the 20th century recorded in oak tree rings." *Journal of Environmental Quality*, **vol. 26**:1638–1643.

[75] Jönsson, P. and Holmquist, B. (1995). "Wind direction in southern Sweden 1740-1992: Variation and correlation with temperature and zonality." *Theoretical Applied Climatology*, **vol. 51**:183–198.

[76] Kabata-Pendias, A. and Pendias, H. (2001). Trace Elements in Soils and Plants. CRC Press, New York.

[77] Keys, D. (1999). Catastrophe. Century, London.

[78] Kindbom, K., Svensson, A., Sjoberg, K. and Karlsson, G. P. (2001). Trends in air concentration and deposition at background monitoring sites in Sweden - major inorganic compounds, heavy metals and ozone. IVL Swedish Environmental Reasearch Institute Report B1429, Göteborg.

[79] Kuniholm, P. E., Kromer, B., Manning, S. W., Newton, M., Latini, C. E. and Bruce, M. J. (1996). "Anatolian tree rings and the absolute chronology of the Eastern Mediterranean 2220-718BC." *Nature*, **vol. 381**:780–782.

[80] LaMarche, V. C. and Hirschboek, K. K. (1984). "Frost rings in trees as records of major volcanic eruptions." *Nature*, **vol. 307**:121–126.

[81] Latimer, S. D., Devall, M. S., Thomas, C., Ellgaard, E. G., Kumar, S. D. and Thien, L. B. (1996). "Heavy metals in the environment - dendrochronology and heavy metal deposition in tree rings of baldcypress." *Journal of Environmental Quality*, **vol. 25**:1411–1419.

[82] Lawrence, G. B., David, M. B. and Shortle, W. C. (1995). "A new mechanism for calcium loss in forest-floor soils." *Nature*, **vol. 378**:162–164.

[83] Leavitt, S. W. and Long, A. (1983). "On a 50 year "climate-free" delta 13c record from juniper tree rings." *Radiocarbon*, **vol. 25**:267–268.

[84] Legge, A. H., Kaufman, H. C. and Winchester, J. W. (1984). "Tree-ring analysis by PIXE for a historical record of soil chemistry response to acidic air pollution." *Nuclear Instruments and Methods in Physics Research*, **vol. 3**:507–510.

[85] Legrand, M., Hammer, C., De Angelis, M., Savarino, J., Delmas, R., Clausen, H. and Johnsen, S. J. (1997). "Sulfur-containing species (methanesulfonate and SO_4) over the last climatic cycle in the Greenland Ice Core Project (central Greenland) ice core." *Journal of Geophysical Research*, vol. **102**:26663–26679.

[86] Lepp, N. W. and J, D. G. (1974). "Studies on lateral movement of 210Pb in woody stems." *Oecologia*, vol. **16**:179–184.

[87] Leuschner, H. H., Sass-Klaassen, U., Jansma, E., Baillie, M. G. L. and Spurk, M. (2002). "Subfossil European bog oaks: population dynamics and long-term growth depressions as indicators of changes in the Holocene hydro-regime and climate." *The Holocene*, vol. **12**:695–706.

[88] Lin, Z. Q., Barthakur, N. N., Schuepp, P. H. and Kennedy, G. G. (1995). "Uptake and translocation of Mn and Zn applied on foliage and bark surfaces of balsam fir (*Abies balsamea* (L.) Mill.) seedlings." *Environmental and Experimental Botany*, vol. **35**:475–483.

[89] Long, R. P. and Davis, D. D. (1989). "Major and trace element concentrations in surface organic layers, mineral soil, and white oak xylem downwind from a coal-fired power plant." *Canadian Journal of Forest Research*, vol. **19**:1603–1615.

[90] Lövestam, G., Johansson, E., Johansson, S. and Pallon, J. (1990). "Elemental micro patterns in tree rings - a feasibility study using scanning proton microprobe analysis." *Ambio*, vol. **19**:87–93.

[91] Lukaszewski, Z., Siwecki, R., Opydo, J. and Zembrzuski, W. (1993). "The effect of industrial pollution on copper, lead, zinc and cadmium concentration in xylem rings of resistant (*Populus marilandicia*) and sensitive (*P. balsamifera*) species of poplar." *Trees*, vol. **7**:169–174.

[92] Maclauchlan, L. E., Borden, J. H., Cackette, M. R. and M, D. J. (1987). "A rapid multisample technique for the detection of trace elements in trees by energy-dispersive X-ray flourescence." *Canadian Journal of Forest Research*, vol. **17**:1124–1130.

[93] Maclauchlan, L. E., Borden, J. H., Cackette, R., M and D'Auria, J. M. (1987). "A rapid, multisample technique for detection of trace elements in trees by energy-dispersive X-ray flourescence spectroscopy." *Canadian Journal of Forest Research*, vol. **17**:1124–1130.

[94] Manning, S. W. (1998). "Correction. new GISP2 ice-core evidence supports 17th century BC date for the Santorini (Minoan) eruption: response to Zielinski and Germani (1998)." *Journal of Archaeological Science*, vol. **25**:1039–1042.

[95] Manning, S. W. (1999). A Test Of Time: The volcano of Thera and the chronology and history of the Aegean and east Mediterranean in the mid second millennium BC. Oxbow books, Oxford.

[96] Manning, S. W., Kromer, B., Kuniholm, P. I. and Newton, M. W. (2001). "Anatolian tree rings and a new chronology for the east Mediterranean Bronze-Iron ages." *Science*, vol. **294**:2494–2495.

[97] Martin, R. R., Zanin, J. P., Bensette, M. J., Lee, M. and Furimsky, E. (1997). "Metals in the annual rings of eastern white pine (*Pinus strobus*) in southwestern Ontario by secondary ion mass spectroscopy (sims)." *Canadian Journal of Forest Research*, vol. **27**:76–79.

[98] Matthews, G. (1993). The carbon content of wood. Forestry Commission Technical paper 4, Edinburgh.

[99] McClenahen, J. R., Vimmerstedt, J. P. and Scherzer, A. J. (1989). "Elemental concentrations in tree rings by PIXE: statistical variability, mobility, and effects of altered soil chemistry." *Canadian Journal of Forest Research*, vol. **19**:880–888.

[100] McCormick, M. P., Thomason, L. W. and Trepte, C. R. (1995). "Atmospheric effects of the Mt Pinatubo eruption." *Nature*, **vol. 373**:399–404.

[101] Meisch, H. U., Kessler, M., Reinle, W. and Wagner, A. (1986). "Distribution of metals in annual rings of the beech (*Fagus sylvatica*) as an expression of environmental changes." *Experientia*, **vol. 42**:537–541.

[102] Meyer, E., Sarna-Wojcicki, A. M., Hillhouse, J. W., Woodward, M. J., Slate, J. L. and Sorg, D. H. (1991). "Fission-track age (400,000 yr) of the Rockland tephra, based on inclusions of zircon grains lacking fossil fission tracks." *Quaternary Research*, **vol. 35**:367–382.

[103] Myre, R. and Camiré, C. (1994). "Distribution of P, K, Ca, Mg, Mn and Zn in the stem of European larch and tamarack." *Annales des Sciences Forestieres*, **vol. 51**:121–134.

[104] Newhall, C. G. and Self, S. (1982). "The volcanic explosivity index (VEI): an estimate of explosive magnitude for historical volcanism." *Journal of Geophysical Research (Oceans and Atmospheres)*, **vol. 87**:1231–1238.

[105] Newnham, R. E., Eden, D. N., Lowe, D. J. and Hendy, C. H. (2003). "Rerewhakaaitu tephra, a land-sea marker for the last termination in New Zealand, with implications for global climate change." *Quaternary Science Reviews*, **vol. 22**:289–308.

[106] Nohrstedt, H. (2002). "Effects of liming and fertilisation (N,PK) on chemistry and nitrogen turnover in acidic forest soils in SW Sweden." *Water, Air and Soil Pollution*, **vol. 139**:343–354.

[107] Oeschger, H. and Langway, C. C. (1989). *The Environmental Record In Glaciers and Ice Sheets*. John Wiley & Sons, Chichester.

[108] Okada, N., Katayama, Y., Nobuchi, T., Ishimaru, Y. and Aoki, A. (1993). "Trace elements in the stems of trees vi - comparisons of radial distributions among hardwood stems." *Mokuzai Gakkaishi*, **vol. 39**:1119–1127.

[109] Okuno, M. and Nakamura, T. (2003). "Radiocarbon dating of tephra layers: Recent progress in Japan." *Quaternary International*, **vol. 105**:49–56.

[110] Oliveira, H., Fernandes, E. A. N. and Ferraz, E. (1997). "Determination of trace elements in tree rings of *Pinus* by neutron activation analysis." *Journal of Radioanalytical and Nuclear Chemistry*, **vol. 217**:125–129.

[111] Oppenheimer, C. (2002). "Limited global change due to the largest known Quaternary eruption, Toba 74kyr BP." *Quaternary Science Reviews*, **vol. 21**:1593–1609.

[112] Oppenheimer, C. (2003). "Climate, environmental and human consequences of the largest known historic eruption: Tambora volcano (Indonesia) 1815." *Progress in Physical Geography*, **vol. 27**:230–259.

[113] Orlandi, M., Pelfini, M., Pavan, M., Santilli, M. and Colombini, M. P. (2002). "Heavy metals variations in some conifers in Valle d'Aosta (western Italian Alps) from 1930 to 2000." *Microchemical Journal*, **vol. 73**:237–244.

[114] Padilla, K. L. and Anderson, K. A. (2002). "Trace element concentration in tree-rings biomonitoring centuries of environmental change." *Chemosphere*, **vol. 49**:575–585.

[115] Palais, J. M., Germani, M. S. and Zielinski, G. A. (1992). "Inter-hemispheric transport of volcanic ash from a 1259 AD volcanic eruption to the Greenland and Antarctic icesheets." *Geophysical Research Letters*, **vol. 19**:801–804.

[116] Patris, N., Delmas, R. J. and Jouzel, J. (2000). "Isotopic signatures of sulfur in shallow Antarctic ice cores." *Journal of Geophysical Research*, **vol. 105**:7071–7078.

[117] Pearce, N. J. G., Perkins, W. T., Westgate, J. A., Gorton, M. P., Jackson, S. E., Neal, C. R. and Chenery, S. P. (1996). "A compilation of new and published major and trace element data for NIST SRM 612 glass reference materials." *Geostandards Newsletter: Journal of Geostandards and Geoanalysis*, **vol. 31**:115–144.

[118] Penninckx, V., Glineur, S., Gruber, W., Herbauts, J. and Meerts, P. (2001). "Radial variations in wood mineral element concentrations: a comparison of beech and pedunculate oak from the Belgian Ardennes." *Annals of Forest Science*, **vol. 58**:253–260.

[119] Percy, K. E. and Baker, E. A. (1988). "Effects of simulated acid rain on leaf wettability, rain retention and uptake of some inorganic ions." *New Phytologist*, **vol. 108**:75–82.

[120] Persson, C., Langner, J. and Robertson, L. (1995). "Dispersion and deposition of air pollutants over Sweden - assessments based on the match model and air pollution measurements." *Papers for the World Clean Air Congress - Atmospheric Pollution*, **vol. 2**:309.

[121] Pillay, K. K. S. (1976). "Activation analysis and dendrochronology for estimating pollution histories." *Journal of Radioanalytical Chemistry*, **vol. 32**:151–171.

[122] Prohaska, T., Stadlbauer, C., Wimmer, R., Stingeder, G., Latkoczy, C., Hoffmann, E. and Stephanowitz, H. (1998). "Investigation of element variability in tree rings of young Norway spruce by laser-ablation-ICPMS." *The Science of the Total Environment*, **vol. 219**:29–39.

[123] Pyle, D. M. (1989). "Ice-core acidity peaks, retarded tree growth and putative eruptions." *Archaeometry*, **vol. 31**:88–91.

[124] Ragsdale, H. L. and Berish, C. W. (1988). "The decline of lead in tree rings of *Carya* spp. in urban Atlanta, GA, USA." *Biogeochemistry*, **vol. 6**:21–29.

[125] Rampino, M. R. and H, A. S. (2000). "Volcanic winter in the Garden of Eden: The Toba supereruption and the late Pleistocene human population crash." *Special Papers; The Geological Society of America*, **vol. 345**:71–82.

[126] Rampino, M. R. and Self, S. (1982). "Historic eruptions of Tambora (1815), Krakatau (1883) and Agung (1963), their stratospheric aerosols and climatic impact." *Quaternary Research*, **vol. 18**:127–143.

[127] Rampino, M. R. and Self, S. (1993). "Climate-volcanism feedback and the Toba eruption of 74,000 years ago." *Quaternary Research*, **vol. 40**:269–280.

[128] Rampino, M. R., Self, S. and Stothers, R. B. (1988). "Volcanic winters." *Annual Review of Earth and Planetary Science.*, **vol. 16**:73–99.

[129] Robitaille, G. (1981). "Heavy-metal accumulation in the annual rings of balsam fir *Abies basamea* (L.) Mill." *Environmental Pollution*, **vol. 2**:193–201.

[130] Robock, A. (2002). "The climatic aftermath." *Science*, **vol. 295**:1242–1244.

[131] Ruhling, A. and Tyler, G. (2001). "Changes in atmospheric deposition rates of heavey metals in Sweden. a summary of nationwide Swedish surveys in 1968/70 - 1995." *Water, Air and Soil Pollution*, **vol. 1**:311–323.

[132] Saka, S. and Goring, D. (1983). "The distribution of inorganic constituents in black spruce wood as determined by TEM-EDXA." *Mokuzai Gakkaishi*, **vol. 29**:648–656.

[133] Sato, S., Doi, T. and Sato, J. (1999). "A temporal increase in the atmospheric 210Pb concentration possibly due to the 1991 eruption of Pinatubo volcano." *Radioisotopes*, **vol. 48**:522–529.

[134] Schweingruber, F. H. (1996). Tree rings and environment dendroecology. Paul Haupt Publishers, Berne.

[135] Scuderi, L. A. (1990). "Tree-ring evidence for climatically effective volcanic eruptions." *Quaternary Research*, **vol. 34**:67–85.

[136] Sear, C. B., Kelly, P. M., Jones, P. D. and Goodess, C. M. (1987). "Global surface-temperatures responses to major volcanic eruptions." *Nature*, **vol. 330**:365–367.

[137] Self, R. and Sparks, R. S. J. e. (1981). Tephra studies. Kluwer Academic Publishers, Dordrecht.

[138] Selin, E., Standzenieks, J., Bowman, J. and Teeyasoontranont, V. (1993). "Multi-element analysis of tree rings by EDXRF spectrometry." *Journal of X-ray Spectrometry*, **vol. 22**:281–285.

[139] Shane, P., Smith, V. and Nairn, I. (2003). "Biotite composition as a tool for the identification of Quaternary tephra beds." *Quaternary Research*, **vol. 59**:262–270.

[140] Shortle, W. C. and Bondietti, E. A. (1992). "Timing, magnitude, and impact of acidic deposition on sensitive forest sites." *Water, Air and Soil Pollution*, **vol. 61**:253–267.

[141] Shortle, W. C., Smith, K. T., Minocha, R., Lawrence, G. B. and David, M. B. (1997). "Acidic deposition, cation mobilization and biochemical indicators of stress in healthy red spruce." *Journal of Environmental Quality*, **vol. 26**:871–876.

[142] Simkin, T. and Siebert, L. (1994). Volcanoes of the world, second edition, A regional directory, gazetteer, and chronology of volcanism during the last 10,000 years. Geoscience Press, Tucson.

[143] Smith, K. T. and Shortle, W. C. (1996). "Tree biology and dendrochemistry." In: "Tree rings, environment and humanity,", editors Dean, J. S., Meko, D. M. and Swetnam, T. W., pages 629–634. Radiocarbon, Department of Geosciences, Tucson, Arizona.

[144] Smith, R. B. and Siegel, L. J. (2000). Windows into the Earth. Oxford University Press, Oxford.

[145] Sparks, R. S. J., Wilson, L. and Sigurdsson, H. (1981). "The pyroclastic deposits of the 1875 eruption of Askja, Iceland." *Philosophical Transactions of the Royal Society of London*, **vol. 299**:241–273.

[146] Spiecker, H. (1991). "Liming, nitrogen and phosphorus fertilization and the annual volume increment of Norway spruce stands on long-term permanent plots in Southwestern Germany." *Fertilizer Research*, **vol. 27**:87–93.

[147] Stewart, C., Norton, D. A. and Fergusson, J. E. (1991). "Historical monitoring of heavy metals in kahikatea ring wood in Christchurch, New Zealand." *The Science of the Total Environment*, **vol. 105**:171–190.

[148] Stewart, C. M. (1966). "Excretion and heartwood formation in living trees." *Science*, **vol. 153**:1068–1074.

[149] Stothers, R. B. and Rampino, M. R. (1983). "Historic volcanism, European dry fogs and Greenland acid preciptation, 1500BC to AD1500." *Science*, **vol. 222**:411–412.

[150] Sunden, A., Brelid, H., Rindby, A. and Engstrom, P. (2000). "Spatial distribution and modes of chemical attachment of metal ions in spruce wood." *Journal of Pulp and Paper Science*, **vol. 26**:352–357.

[151] Symeonides, C. (1979). "Tree-ring analysis for tracing the history of pollution: application to a study in nothern Sweden." *Journal of Environmental Quality*, **vol. 8**:482–486.

[152] Szopa, P. S., McGinnes, E. A. and Pierce, J. O. (1973). "Distribution of lead within the xylem of trees exposed to air-borne lead compounds." *Wood Science*, **vol. 6**:72–77.

[153] Tendel, J. and Wolf, K. (1988). "Distribution of nutrients and trace elements in annual rings of pine trees (*Pinus silvestris*) as an indicator of environmental changes." *Experientia*, **vol. 44**:975–980.

[154] Thomas, C. E., Latimer, S. D., Mills, O. P. and Le, K. H. N. (1996). "Dendrochemistry of loblolly pine and cypress cores: initial results for forest health monitoring." In: "Tree rings, environment and humanity,", editors Dean, J. S., Meko, D. M. and Swetnam, T. W. Radiocarbon, Department of Geosciences, Tucson, Arizona.

[155] Thorarinsson, S. (1981). "Greetings from Iceland. Ash-falls and volcanic aerosols in Scandinavia." *Geografisker Annaler*, **vol. 63**:109–118.

[156] Tomita, M., Katayama, Y., Nishimura, K. and Takada, J. (1990). "Radial distribution of Sb and Cu in the stem of woody plants grown on the Sb-polluted soil as well as on the Cu-enriched soil near an ore dressing house of copper mine." *Radioisotopes*, **vol. 39**:553–556.

[157] Tommasini, S., Davies, G. R. and Elliott, T. (2000). "Lead isotope composition of tree rings as bio-geochemical tracers of heavy metal pollution: A reconnaissance study from Firenze, Italy." *Applied Geochemistry*, **vol. 15**:891–900.

[158] Ward, N. I., Brooks, R. R. and Reeves, R. D. (1974). "Effect of lead from motor-vehicle exhausts on trees along a major thoroughfare in Palmerston North, New Zealand." *Environmental Pollution*, **vol. 6**:149–158.

[159] Wardell, J. F. and Hart, J. H. (1973). "Radial gradients of elements in white oak wood." *Wood Science*, **vol. 5**:298–303.

[160] Watmough, S. A. (1999). "Monitoring historical changes in soil and atmospheric trace metal levels by dendrochemical analysis." *Environmental Pollution*, **vol. 106**:391–403.

[161] Watmough, S. A. (2002). "A dendrochemical survey of sugar maple (*Acer saccharum* Marsh.) in south central Ontario, Canada." *Water, Air and Soil Pollution*, **vol. 136**:165–187.

[162] Watmough, S. A. and C, H. T. (2002). "Historical changes in lead concentrations in tree-rings of sycamore, oak and scots pine in north-west England." *The Science of the Total Environment*, **vol. 293**:85–96.

[163] Watmough, S. A. and Hutchinson, T. C. (1996). "Analysis of tree rings using inductively coupled plasma mass spectrometry to record fluctuations in a metal pollution episode." *Journal of Environmental Pollution*, **vol. 93**:93–102.

[164] Watmough, S. A. and Hutchinson, T. C. (1999). "Change in the dendrochemistry of sacred fir close to Mexico City over the past 100 years." *Environmental Pollution*, **vol. 104**:79–88.

[165] Watmough, S. A., Hutchinson, T. C. and Evans, R. D. (1997). "Application of laser ablation inductively coupled plasma mass spectrometry in dendrochemical analysis." *Environmental Science and Technology*, **vol. 31**:114–118.

[166] Watmough, S. A., Hutchinson, T. C. and Evans, R. D. (1998). "Development of solid calibration standards for trace elemental analyses of tree rings by laser ablation inductively coupled plasma-mass spectrometry." *Environmental Science and Technology*, **vol. 32**:2185–2190.

[167] Watmough, S. A., Hutchinson, T. C. and Evans, R. D. (1998). "The quantitative analysis of sugar maple tree rings by laser ablation in conjunction ICP-MS." *Journal of Environmental Quality*, **vol. 27**:1087–1094.

[168] Wimmer, R. and McLaughlin, S. B. (1996). "Possible relationships between chemistry and mechanical properties in the microstructure of red spruce xylem." In: "Tree rings, environment and humanity,", editors Dean, J. S., Meko, D. M. and Swetnam, T. W. Radiocarbon, Department of Geosciences, Tucson, Arizona.

[169] Yadav, R. R. (1993). "Reply to Yamaguchi's letter to the editor on tree-ring records of volcanic events in Kamachatka and elsewhere." *Quaternary Research*, **vol. 40**:264.

[170] Yin, X., Foster, N. W., Morrison, I. K., and Arp, P. A. (1994). "Tree-ring-based growth analysis for a sugar maple stand: relations to local climate and transient soil properties." *Canadian Journal of Forest Research*, **vol. 24**:1567–1547.

[171] Zayed, S., J Loranger and Kennedy, G. (1992). "Variations of trace element concentrations in red spruce tree rings." *Water, Air and Soil Pollution*, **vol. 65**:281–291.

[172] Zielinski, G. A. and Germani, M. S. (1998). "New ice-core evidence challenges the 1620's BC age for the Santorini (Minoan) eruption." *Journal of Archaeological Science*, **vol. 25**:279–289.

[173] Zielinski, G. A. and Germani, M. S. (1998). "Reply to: Correction. new GISP2 ice-core evidence supports 17th century BC date for the Santorini (Minoan) eruption." *Journal of Archaeological Science*, **vol. 25**:1043–1045.

[174] Zielinski, G. A., Mayewski, P. A., Meeker, L. D., Whitlow, S. and Twickler, M. S. (1996). "Potential atmospheric impact of the Toba mega-eruption." *Geophysical Research Letters*, **vol. 23**:837–840.

[175] Zillén, L. M., Wastegardrd, S., and Snowball, I. F. (2002). "Calendar year ages of three mid-Holocene tephra layers identified in varved lake sediments in west central Sweden." *Quaternary Science Reviews*, **vol. 21**:1583–1591.

Appendix A

Tree species - common names

TABLE A.1: A reference table of tree species occurring in the text, full and common names. The full names are arranged alphabetically to facilitate use. Where 'sp.', only the genus is known.

Full name	Common Name
Abies balsamea (L.) Mill.	Balsam Fir
Abies religiosa H.B.K. & Cham.	Sacred Fir
Acanthopanax sciadophylloides Fr. & Sav.	Koshiabura
Acer sp.	Maple
Acer pseudoplatanus L.	Sycamore
Acer rubrum L.	Red maple
Acer saccharum Marsh.	Sugar maple
Acer sieboldianum Miq.	Kohauchiwakaede
Aesculus hippocastanum L.	Horse chestnut
Aesculus turbinata Blume.	Tochinoki
Agathis austrailis (D.Don) Salisb.	Kauri
Alnus glulinosa L.	Alder
Alnus rubra Bong.	Red alder
Betula grossa Siebold & Zucc.	Mizume
Betula pubescens L.	Downy birch
Betula sp.	Birch
Carya sp.	Hickory
Castanea crenata Siebold & Zucc.	Kuri
Celtis australis L.	European Hackberry
Celtis sinensis Nakai	Enoki
Clethra barbinervis Siebold & Zucc.	Ryoubu
Cryptomeria japonica D. Don.	Sugi
Dacrycarpus dacrydioides	Kahikatea (White Pine)
Diospros kaki Thunb.	Kaki
Fagus crenata Blume	Buna
Fagus sylvatica L.	Beech
Fraxinus excelsior L.	Ash
Hovenia dulcis Thunb.	Kenponashi
Juniperus communis L.	Common Juniper
Juniperus virginiana L.	Eastern red cedar
Kalopanax pictus Nakai.	Harigiri

APPENDIX A. TREE SPECIES - COMMON NAMES

Larix decidua Mill.	Larch
Lindera erythrocarpa Makino.	Kanakuginoki
Liriodendron tulipifera L.	Yurinoki
Magnolia obovata Thunb.	Hoonoki
Phellodendron amurense Rupr.	Kihada
Picea abies (L.) H.Karst.	Norway spruce
Picea glauca	White spruce
Picea mariana (Mill.) Britton, Sterns & Poggenb.	Spruce
Picea rubens Sarg.	Red Spruce
Picea sitchensis (Bong.) Carr.	Sitka Spruce
Picea sp.	Spruce
Pinus aristata Engelm.	Bristlecone Pine
Pinus banksiana Lamb.	Jack Pine
Pinus contorta x banksiana (Loud.) Lamb.	Lodgepole-jack pine
Pinus contorta Douglas and Loudon.	Lodgepole pine
Pinus densiflora Siebold & Zucc.	Akamatsu
Pinus echinata Mill.	Short-leaf pine
Pinus elliottii	Slash pine
Pinus halepensis Mill.	Aleppo pine
Pinus koraiensis Siebold & Zucc.	Korean pine
Pinus nigra Ait.	Black pine
Pinus ponderosa Dougl.	Mammoth ponderosa pine
Pinus radiata D.Don.	Monteray pine
Pinus sp.	Pine
Pinus strobus L.	White pine
Pinus sylvestris L.	Scots Pine
Pinus taeda L.	Loblolly Pine
Platanus orientalis L	Oriental Plane
Populus balsamifera L.	Balsam Poplar
Populus marilandica Bosc & Poir.	Poplar
Pseudotsuga menziesii (Mirb.) Franco	Douglas fir
Quercus alba L.	White oak
Quercus mongolica Fisch. & Turcz.	Mongolian oak
Quercus nigra L.	Water oak
Quercus prinus L.	Chestnut oak
Quercus robur L.	Common oak
Quercus rubra L.	Red oak
Quercus serrata Thunb.	Konara
Quercus sp.	Oak
Sequoiadendron giganteum (Lindl.) Buchholz.	Giant Sequoia
Sorbus alnifolia K. Koch.	Azukinashi
Sorbus commixta Hedl.	Nanakamado
Stewartia pseudo-camellia Maxim.	Natsutsubaki
Styrax obassia Siebold & Zucc.	Hakuunboku
Taxodium distichum (L.) Rich.	Baldcypress
Tilia europaea L.	Lime
Tsuga canadensis (L.) Carr.	Eastern hemlock
Ulmus sp.	Elm
Ulmus glabra Huds.	Wych elm
Ulmus procera L.	English elm
Zelkova serrata Makino.	Keyaki

Appendix B

Element names and symbols

TABLE B.1: A reference table of element symbols and names referred to in the text. The symbols are arranged alphabetically to facilitate use.

Symbol	Element	Symbol	Element
Ag	Silver	Mn	Manganese
Al	Aluminum	Mo	Molybdenum
Ar	Argon	N	Nitrogen
As	Arsenic	Na	Sodium
Au	Gold	Nb	Niobium
B	Boron	Nd	Neodymium
Ba	Barium	Ne	Neon
Be	Beryllium	Ni	Nickel
Bi	Bismuth	O	Oxygen
Br	Bromine	P	Phosphorus
C	Carbon	Pb	Lead
Ca	Calcium	Pd	Palladium
Cd	Cadmium	Pm	Promethium
Ce	Cerium	Pr	Praseodymium
Cl	Chlorine	Pt	Platinum
Co	Cobalt	Ra	Radium
Cr	Chromium	Rb	Rubidium
Cs	Cesium	Re	Rhenium
Cu	Copper	Rh	Rhodium
Dy	Dysprosium	Rn	Radon
Er	Erbium	S	Sulphur
Eu	Europium	Sb	Antimony
F	Flourine	Sc	Scandium
Fe	Iron	Se	Selenium
Ga	Gallium	Si	Silicon
Gd	Gadolinium	Sm	Samarium
Ge	Germanium	Sn	Tin
H	Hydrogen	Sr	Strontium
He	Helium	Ta	Tantalum
Hf	Hafnium	Tb	Terbium
Hg	Mercury	Tc	Technetium

APPENDIX B. ELEMENT NAMES AND SYMBOLS

Ho	Holmium	Te	Tellurium
Hs	Hassium	Th	Thorium
I	Iodine	Ti	Titanium
In	Indium	Tl	Thalium
Ir	Iridium	Tm	Thulium
K	Potassium	U	Uranium
Kr	Krypton	V	Vanadium
La	Lanthanum	W	Tungsten
Li	Lithium	Xi	Xenon
Lu	Lutetium	Yb	Ytterbium
Md	Mendelevium	Zn	Zinc
Mg	Magnesium	Zr	Zirconium

Appendix C

CEM Digestion of Wood

This method was adapted from the CEM Microwave sample preparation note for wood pulp. Equipment required was a CEM Microwave Sample Preparation System and lined digestion vessel accessory set (6 vessels). The reagent used was ultrapure 70% Nitric acid.

1. 0.5g of chipped sample was weighed into digestion vessels. 10ml of Nitric acid was added to each vessel.

2. All vessels, except the one to be used as a pressure control, were sealed.

3. The control vessel was sealed with a modified cap assembly.

4. The vessels were placed into the turn table. The vents were connected via the vent tubes from the vessels to the collection vessel.

5. The turntable was placed into the system. The pressure sensing line was connected to the control vessel.

6. The microwave was programmed according to the settings shown in table C.1, and the program was run to completion.

7. The samples were cooled for 10 minutes, the control vessel was vented and the pressure sensing line removed. The whole turntable was removed from the system.

8. The vessels were manually vented and then opened. The solutions were transferred into sterile sampling tubes for dilution and analysis.

TABLE C.1: Microwave digestion for wood pulp, programming information for a nominal 630 watt system.

Stage	1	2	3	4	5
%power	52.8	52.8	52.8	52.8	52.8
PSI	40	85	135	175	0
Time	10:00	10:00	10:00	10:00	0:00
TAP	5.00	5.00	5.00	5.00	0.00
Fan Speed	100	100	100	100	100

Appendix D

Set-up for LA-ICP-MS analytical conditions

These analytical conditions were set for all LA-ICP-MS analysis and are particular to the system used at the NERC ICP-MS Lab at Kingston-Upon-Thames, in conjunction with the analysis of wood samples.

TABLE D.1: ICP-MS details

Instrument	Plasma Quad 2 plus STE, ThermoElemental
Forward power	1349 W
Reflected power	2 W
Nebulizer gas flow	0.982 l/min
Auxiliary gas flow	1.7 l/min
Coolant gas flow	14 l/minP

TABLE D.2: Data acquisition details

Mode	Scanning / peak jumping
Scan regions	12 - 24 - 27 - 52 - 66
Dwell time	300
Channels per amu	20
Total acquisition time	25 sec
Repeats	3
Calibration method	External

TABLE D.3: Laser details

Type	Cetac LSX-100
Wavelength	266 nm
Repetition rate	10 shots
Analysis mode	scanning
Pulse energy (Level, mJ)	Level 5 (0.3 - 0.5 mJ)
Laser focusing	$500\mu m$
Scan Speed	$10\mu m$ / s
Distance between lines	$50\mu m$

Appendix E

Method for soils into solutions

Adapted from Reading Geochemistry notes, procedure digest 05: total decomposition for the determination of trace elements. Equipment required: PTFE 30ml beakers, glass rods, a hotplate, an acid *washdown* fume cupboard, plastic funnels, HDPE sample bottles, Whatman 542 filter papers, an Oxford pipette and tips, 50ml volumetric flasks, silver crucibles and a furnace. Reagents required were: 48% Hydrofluoric acid, 60% Perchloric acid, 69% Nitric acid, grade 1 Hydrochloric acid, UHQ water and Sodium Hydroxide pellets. The procedure was as follows:

1. 0.005g sample was weighed into a PTFE beaker. Using an Oxford pipette, 5ml of Nitric acid was to each beaker and swirled gently.

2. The hotplate was set up in the fume cupboard up at a temperature of 160°C, beakers were placed on it and left for six hours.

3. Using tongs, the lid was carefully removed and the sample evaporated to dryness.

4. Using the Oxford pipette, 5ml of Nitric acid and 5mls of Perchloric acid were added to each beaker and swirled gently.

5. 10ml of Hydrofluoric acid were measured out with a plastic measuring cylinder and added to each beaker.

6. Beakers were placed on the hotplate at 200°C and left overnight.

7. Beakers were removed from the heat and allowed to cool in the fume cupboard.

8. A further 2ml of Perchloric acid were added to each beaker with the pipette, and the beakers were placed back on the hotplate and evaporated to near dryness.

9. The beakers were removed from the heat and allowed cool in the fume cupboard.

10. Using the pipette, 5ml of Hydrochloric acid were added to each beaker, swirled and the beakers were replaced on the hotplate at 100°C.

11. Once all the residual rock had dissolved, the solutions were cooled and filtered through the ashless filter papers to be collected in a 100ml glass beaker.

12. The filter papers were placed in silver crucibles, transferred to a tray and taken to the main lab.

13. The papers were placed in a cold furnace. The furnace was switched on and the temperature set at 800°C, the samples were left in the furnace at this temperature for 30 minutes.

APPENDIX E. METHOD FOR SOILS INTO SOLUTIONS

14. The samples were removed from the furnace and left to cool before 5 Sodium Hydroxide pellets were added to each crucible, and the samples were placed back in the furnace at 800°C for 30 minutes.

15. The samples were removed from the furnace, swirled till the remaining mixture solidified and left to cool.

16. 5ml of ultrapure deionised water was added to each crucible and left to dissolve. 3 ml of hydrochloric acid were added and the mixture was stirred.

17. Each dissolved, neutralised sample was then added to its corresponding filtrate and made up to 50mls with water in a volumetric flask.

18. The finished samples were transferred to labelled sample tubes and cold stored till analysis.

www.ingramcontent.com/pod-product-compliance
Ingram Content Group UK Ltd.
Pitfield, Milton Keynes, MK11 3LW, UK
UKHW060200240426
12048UKWH00029B/1669